T0199258

Transport and Interactions of Chlorides in Cement-Based Materials

Transport and Interactions of Chlorides in Cement-Based Materials

Caijun Shi, Qiang Yuan,
Fuqiang He, and Xiang Hu

CRC Press
Taylor & Francis Group
Boca Raton London New York

CRC Press is an imprint of the
Taylor & Francis Group, an **informa** business

CRC Press
Taylor & Francis Group
6000 Broken Sound Parkway NW, Suite 300
Boca Raton, FL 33487-2742

First issued in paperback 2021

© 2020 by Taylor & Francis Group, LLC
CRC Press is an imprint of Taylor & Francis Group, an Informa business

No claim to original U.S. Government works

ISBN 13: 978-1-03-209096-2 (pbk)
ISBN 13: 978-1-138-49270-7 (hbk)

Visit the Taylor & Francis Web site at
http://www.taylorandfrancis.com

and the CRC Press Web site at
http://www.crcpress.com

Publisher's Note
The publisher has gone to great lengths to ensure the quality of this reprint but points out that some imperfections in the original copies may be apparent.

Contents

Preface

Chloride-induced corrosion of steel bar in concrete ranks number one in durability issues of reinforced concrete structures. Understanding the transport and interactions of chlorides in concrete is the first step in controlling the chloride-induced corrosion of reinforced concrete. Due to extensive research funded by governments or industries around the world, a great deal of knowledge has been generated, great advances have been made, and many engineering experiences have been obtained over the past few decades. Durability design has become dominant in the design of important costal infrastructures which may subject to chloride attack. However, this doesn't mean that we have completely understood the transport and interactions of chloride in concrete and can accurately predict and control it. Still, many scientific issues are yet to be clarified.

This book focuses on the chloride transport and interactions in cement and concrete. It reviews the general knowledge, technologies, and experiences involved in these areas, and presents the state-of-art progress made recently. The authors of this book have long been involved in research related to chloride transport and interactions in concrete at Hunan University, Ghent University, Central South University, and Xiamen University of Technology. Over the past few decades, many graduate students have obtained their M.Sc. or Ph.D. degrees in this field of study, and the theses are important sources for this book. Obviously, without the knowledge and experiences obtained by them, without their extraordinary works, this book would not be possible. The authors would like to express their gratitude to each and every one of them who have made contributions to the content of this book. Special thanks go to Dr. Runxiao Zhang, a postdoctoral fellow in the Department of Civil Engineering at the University of Toronto, for his participation in the preparation of Chapter 7, and the students who helped with organizing the format of entire book, as well as drawing the new graphs and tables.

The intended audience of this book includes students, researchers, and practicing engineers in the concrete community. This book can also be used as a reference book or extensive reading book for graduate courses on durability of reinforced concrete structures. Researchers may be inspired by the comprehensive overview on the chloride transport and interaction in cement and concrete. Practicing engineers can also benefit from this book through learning the basic knowledge and practical techniques.

Caijun Shi
Qiang Yuan
Fuqiang He
Xiang Hu

Authors

Caijun Shi is a professor in the College of Building Materials at Hunan University, China. He is author of *Alkali-Activated Cements and Concretes*, also published by CRC Press.

Qiang Yuan is a professor at Central South University, China.

Fuqiang He is a professor at Xiamen University of Technology, China.

Xiang Hu is a PhD student at Hunan University, China.

Chapter 1

Introduction

The deterioration of reinforced concrete structures caused by reinforcement corrosion is a worldwide concrete durability problem, particularly when the concrete structures are located in marine environments. Mostly, concrete constructions can be submitted to attacks from aggressive substances, such as chloride and carbon dioxide, chemically. Among these, chloride-related reinforcing steel corrosion accounts for the most of them. When chloride concentration in the pore solution around steel bars reaches a threshold value and breaks the passivation film, corrosion can be initiated at some connection points of the steel bar until the local damage of concrete. For concrete structures exposed to extreme conditions, such as marine exposition or highway structures where de-icing salt is used, the damage caused by chloride ingress does great harm to the durability of the reinforcement concrete structure and gives rise to high repairing and reconstruction costs.

1.1 CHLORIDE-RELATED CORROSION

Normally, the concrete embedded in concrete is highly resistant to corrosion under the protection of cover layer. In well-designed and properly cured concrete with relatively low water to binder (w/b) ratio, the penetration of aggressive substances such as chloride ions and carbon dioxide can be effectively restricted from concrete surface to steel surface. The condition of high alkalinity (pH>13.5) in pore solution of cement-based materials provides an excellent condition for the passivation of steel (Singh and Singh 2012). In one view, steel can be protected from corrosion by the surface layer film (approximately 10000 Å thick), which generally consists of ferric oxide (Fe_2O_3) and is thought to passivate the steel from corrosion (Shan et al. 2008; Waseda et al. 2006; Tittarelli and Bellezze 2010). It can be concluded that in a properly designed, constructed, and maintained reinforced concrete structure, the problem of steel corrosion should be neglected during the service life. However, these requirements on the design, construction, and maintenance of concrete structure are always not achieved in practice, and the corrosion of steel bar in reinforced structures has become a common cause of the decreased durability and service life of concrete structures (Alexander and Nganga 2014).

Generally, it is considered that two stages are included in process of corrosion, especially for pitting corrosion: initial and propagation stages. The initial stage of corrosion relates to the breakdown of the passivation layer, while the propagation stage of corrosion describes the reaction between steel and electrolytes and the formation of corrosion products. There is general agreement about the mechanism of the propagation stage which is the transformation of electrons from anode to cathode surface of steel and the formation of electrical current in steel. In the propagation stage, corrosion products such as $Fe(OH)_3$, $FeSO_4$, Fe_3O_4, $FeO(OH)$, $HFeOOH$, and $HFeO_2$ will be generated based on different composition and ionic concentration of pore solution (Bazant 1979). In this period, re-passivation of steel surface can occur when chloride concentration in pore solution around the steel surface is reduced. It was reported that sufficient chloride concentration was necessary for the propagation of steel corrosion and to keep the steel surface away from re-passivation (Eichler et al. 2009).

Generally, a passive oxide layer can be formed on the steel surface in alkaline conditions, which is followed by electrochemical deposition of polyaniline (PAni) (Jafarzadeh et al. 2011). The doped PAni layer is in contact with the oxidized surface and can stabilize the oxide ions from dissolution or take part in the corrosion process. It was reported (DeBerry 1985) that a PAni layer could be electrochemically deposited above the passive oxide layer formed on steel surfaces, which provide anodic protection from sulfuric acid corrosion. Even though some papers (Kraljić et al. 2003; Sathiyanarayanan et al. 2008) have shown that no corrosion resistance could be provided, the corrosion protection of PAni and the oxide layer in different types of steel have been revealed by many researches (Chaudhari and Patil 2011; Johansen et al. 2012; Karpakam et al. 2011).

In order to describe the initial corrosion stage, several theories have been proposed (Kuang and Cheng 2014), such as local acidification theory (Galvele et al. 1978; Galvele 1976), depassivation–repassivation theory (Dawson and Ferreira 1986; Richardson and Wood 1970), chemical dissolution theory (Hoar and Jacob 1967), point defect models (PDM) (Chao et al. 1981; Urquidi and Macdonald 1985), chemical–mechanical models (Hoar 1967; Sato 1971), and anion penetration/migration models (Okamoto 1973; Rosenfeld and Marshakov 1964). Among all of these theories, a consensus was obtained that the adsorption of aggressive ions, such as chloride ions, plays an important role in the initial stage of steel corrosion (Angst et al. 2011; Cheng et al. 1999). The depassivation of protection layer on the surface of steel is directly induced by reaching the threshold value of chloride concentration in pore solution in contact with steel surface (Ghods et al. 2012). Bertocci and Ye (1984) reported that the most important role of chloride ions on steel corrosion rested on the increasing possibilities of local breakdown of the passive oxide layer. In some studies (Angst et al. 2011;

Liao et al. 2011), it was found that when chloride ions were introduced into the steel surface, no protective oxide layer could be formed in the anodic side and a chloride-ion film was formed to initiate the corrosion. The roles chloride ions played in steel corrosion can be: increase of iron solubility and conductivity of electrolytes, and dissemination of corrosion products (Lou and Singh 2010). Electronically, the existence of chloride ions accelerates the corrosion initiation by increasing the susceptible sites (Burstein et al. 1993) and decreasing the value of pitting potential (Tang et al. 2014; Xu et al. 2010).

Extensive studies (Otieno et al. 2016a; Tennakoon et al. 2017; Hou et al. 2016; Borg et al. 2018; Borade and Kondraivendhan 2019) have been conducted to investigate the effects of chloride-induced corrosion on durability and performance of concrete structures. It was considered that the corrosion of steel bars due to chloride penetration dominated the durability of concrete structures exposed to extreme condition, such as marine exposition or highway structures where de-icing salt was used. The damage caused by chloride ingress does great harm to the durability of reinforcement concrete structures and gives rise to high repairing and reconstruction costs. The deterioration of reinforced concrete due to chloride-related corrosion can be normally divided into four stages, as shown in Figure 1.1 (Berke et al. 2014; Budelmann et al. 2014): the initial cracking of concrete due to external applied pressure or shrinkage of concrete; the penetration of chloride ions in concrete cover and accumulates around the steel surface; the corrosion of steel which results in the formation and accumulation of corrosion products; the generation of cracks in concrete cover and fracture deterioration. Generally, the permeability of concrete cover and the diffusion rate of chloride ions are dominant factors of the initial stage, while oxide dissolution, moisture condition, and electrical resistance of concrete control the last three stages of steel corrosion.

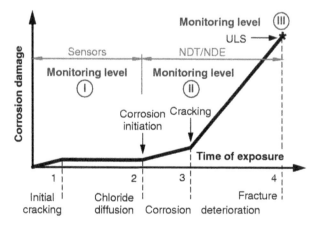

Figure 1.1 Steel corrosion stages in concrete. (From Budelmann et al. 2014.)

As it has become a major problem threatening concrete structures, studies on damage prevention in chloride-induced steel corrosion have been extensively conducted (Ann and Song 2007; Pour-Ali et al. 2015; Van Belleghem et al. 2018). According to the chloride-related corrosion mechanism described previously, efforts that have been made to prevent chloride corrosion involve keeping chloride ions from penetrating into steel surface (Pack et al. 2010; Song et al. 2008) and increasing the chloride binding capacity of cement matrix from the concrete point of view (Glass and Buenfeld 2000; Yuan et al. 2009).

1.2 CHLORIDE TRANSPORT IN CEMENT-BASED MATERIALS

Generally, chloride ion can be introduced into concrete in two ways: (1) as an admixture (internal chloride); (2) penetration from external environment (external chloride), mainly from seawater and de-icing salt. The internal chloride can be well controlled by using chloride-free ingredients and choosing raw materials with low chloride content. But the chloride ions penetrated from the environment are mostly inevitable and difficult to predict and control. The service life of reinforced concrete depends more on chloride ions penetrating from the external environment. Thus, the transport of chloride in concrete has attracted a lot of attention. For the external chloride, no matter from where the chloride externally originated, by de-icing salt or seawater, chloride generally penetrates into concrete with the transportation of water within materials. When chloride ions continue to penetrate into concrete or chloride ingress is repeated, the chloride concentration around the steel surface can be relatively high.

As a homogenous porous material, cement concrete is mostly permeable by chloride ions from the surrounding environment. Based on different driven forces, the transport of chloride ions in cement concrete can be divided into five classification groups (Yuan et al. 2009): hydrostatic advection, capillary suction, diffusion, electrical migration, and thermal migration. The saturation degree of a concrete structure is always the foremost factor controlling the chloride transport process. For some marine constructions, concrete structures may always experience exposure to sea water and be exposed to dry condition periodically. When concrete is exposed to salt water or a high relative humidity environment, the salt water can enter and increase the chloride concentration in pore solution. If the external environment becomes dry with the removal of exposure water or decrease of relative humidity, the internal water of the concrete can evaporate and shift in different directions with the movement under wet conditions. During this process, only water can be evaporated, and salts are left. Based on the duration of the dry process, the water in the area of the concrete surface can be totally removed. Thus, the chloride concentration in this zone can be increased and diffuses toward parts with lower concentration under the concentration gradient forces. Therefore, in dry conditions, the internal water moves outward from the concrete while salts move inward. Thus, in the next cycle of the wet process with salt water, more salts will be brought into samples with water. The chloride concentration profile along the penetration side may first be decreased, then increased. Generally, the wetting process occurs rapidly while longer a period of time for drying is needed. Therefore, the chloride ingress process and penetration rate depend on the length of the wetting and drying processes.

Studies have been done on ion and fluid transport in cement concrete structures with different saturation degrees, and several models have been proposed to describe the chloride ingress process within concrete structures (Jin et al. 2008; Yang et al. 2006; Otieno et al. 2016). By taking into account the diffusion and sorption of chloride ions in concrete, a durability simulator model was designed to simulate the marine environment with wetting–drying cycle (Iqbal and Ishida 2009). Chloride profiles were simulated by the proposed model with the moisture conductivity. In 1931, Richards (1931) firstly studied the mechanism of chloride transport in unsaturated porous solids, and an equation was proposed to describe the water flow under the capillary suction. Based on this work, Samson et al. (2005) developed a model to describe the ion transport in unsaturated cement systems by coupling the ionic and water transport models within materials. Generally, a parameter or model of dynamical moisture within systems is applied to take the water saturation degree into account during the chloride transport studies.

Besides the wetting–drying circles, an external applied electrical field can also accelerate the penetration of chloride salts into concrete, and it has been widely applied in rapid chloride migration tests. Compared to concentration gradient force, the external applied voltage can accelerate the ingress of chloride salts into concrete samples more significantly. According to the rapid chloride migration (RCM) test, where external electrical voltage was applied, it was reported that the obtained chloride migration coefficient was larger than chloride diffusion coefficient obtained from the natural diffusion test. Even though the RCM test has been widely applied to evaluate the chloride penetration resistance of concrete samples, many controversial issues still remain unsolved. Regarding the movement of chloride ions within cement-based materials, parts of them can be fixed by chemical or physical works between chloride ions and hydration products or other solid phases. These parts of chloride ions are defined as bound chloride, and this phenomenon is known as chloride binding. Free chloride ions in pore solution will be reduced due to chloride binding, and the flow of chloride ions with pore solution can be slowed down. However, the external applied voltage can bring some unexpected changes on chloride binding of cement matrix. On the other hand, the shortening of the ingress period due to electrical voltage can also change the content of chloride ions reacted with solid phase in samples.

Some researchers (Krishnakumark 2014; Spiesz and Brouwers 2013; Voinitchi et al. 2008) reported that the adsorption of chloride ions after reaching the steady state was independent of the applied voltage, and the applied voltage mainly affects the free chloride concentration in pore solution under steady state. Numerical simulation on non-steady-state diffusion and RCM tests showed that the free chloride ions in pore solution could be instantaneously combined, and no variation of free and bound chloride content was found between samples after diffusion and migration tests (Spiesz and Brouwers 2012). However, Spiesz and Brouwers (2013) showed that it took seven days or even longer for chloride binding in cement matrix to reach equilibrium in diffusion tests, which was much longer than the testing duration of RCM test. Castellote et al. (1999, 2001) analyzed the chemically bound and free chloride ions of the samples after RCM test using X-ray fluorescence technique and leaching method, respectively. In their studies, the obtained chloride adsorption isotherm was compared with that from diffusion test by Sergi et al. (1992). The applied electrical field suppressed chloride binding at lower free chloride concentration (<97 g/L) while enhanced it at higher concentration. The decreased contact time and altered electrical double layer (EDL) property were considered as reasons to cause these differences.

It is apparent that under concentration gradient, wetting and drying circles, or external applied voltage, the chloride ions may progressively penetrate through the concrete cover layer toward the surface of steel. Then, along with the different depth from the concrete surface, a chloride profile can be established. As shown in Figure 1.2 (Toumi et al. 2007), the total chloride content gradually decreased with the increase of distance

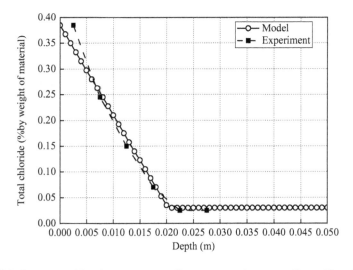

Figure 1.2 Concrete chloride content profile in immersion test. (From Toumi et al. 2007.)

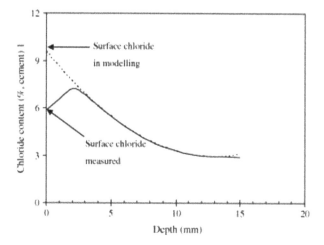

Figure 1.3 A decrease of the surface chloride content at the right surface of concrete in a conventional chloride profile. (From Ann et al. 2009.)

to concrete surface. Sometimes, the chloride concentration in the outmost layer will be reduced, as shown in Figure 1.3 (Ann et al. 2009). There is a sink of the chloride profile in the outermost layer of concrete. The rapid water movement during wetting and drying circle and the decrease of bound chloride content due to the portlandite precipitation were considered as the reasons for this phenomenon (Ann et al. 2009). Generally, the chloride content results of the surface layer are ignored and mathematical fitting is applied to calculate the surface chloride content, as shown in Figure 1.3. Besides the total chloride content, the chloride profile for free chloride and chemically or physically bound chloride content have been also studied in the literature (Glass and Buenfeld 2000; Ishida et al. 2009).

1.3 INTERACTION OF CHLORIDE WITH HYDRATION PRODUCTS OF CEMENT

For concrete structures, chloride ions may derive from aggregates/mixing water (internal chloride) or penetrate into steel surface during exposure to chloride-bearing environments (intruded chloride). During the ingress process into cement concrete, the chloride ions in pore solution can be captured by the solid phases, including unhydrated cement component and hydration products of cementitious materials. The interaction between chloride ions and cement hydrates can be chemical or physical, and are defined as chemical binding and physical adsorption respectively. It was generally considered that the free-state chloride ions in pore solutions were mainly responsible for the corrosion of reinforcement steel (Yuan et al. 2009). However, Glass and Buenfeld (1997, 2000) stated that the bound chloride ions could also induce the corrosion of reinforcement steel when they were released into pore solutions under certain conditions. The effects of chloride binding on chloride penetration and chloride-related corrosion can be considered as three aspects: (1) reduction of the free chloride concentration in the vicinity of the reinforcing steel, which will reduce the chance of corrosion; (2) removal of chloride from the diffusion flux, thus retarding the penetration of chloride to the level of the steel (Li et al. 2015); (3) formation of Friedel's salt, which results in a less porous structure and slows down the transport of chloride ion. Therefore, the effect of chloride binding must be taken into account when studying chloride ion transport in concrete. It can be seen from Figure 1.4 that the consideration of chloride binding can change the chloride profile (Martın-Pérez et al. 2000). However, it was also reported that chloride binding could have no remarkable effects on penetration depth (Ye et al. 2016).

Due to the retardation effect of bound chloride, the free and bound chlorides must be distinguished from each other in service life prediction models. Chemical binding is generally considered as the result of reaction between chlorides and C_3A or AFm phases to form Friedel's salt or the reaction with C_4AF to form a Friedel's salt analog (Florea and Brouwers 2012; Ipavec et al. 2013; Yuan et al. 2009). Physical binding is due to the adsorption of chloride ion to the C–S–H surfaces. Studies on mechanism of chloride binding of cement-based materials mostly focus on C-S-H (Shi et al. 2017) and AFm phase (Chen et al. 2015), as the former controls the physical adsorption while the latter dominates the chemical binding. However, the portlandite and ettringite (Ekolu et al. 2006; Hirao et al. 2005), also the Friedel's salt (Elakneswaran et al. 2009), formed through the interaction of other AFm phases and intruding chlorides can bind chloride ions (Florea and Brouwers 2012). Besides, the alumina phase in mineral component of cementitious materials such as C_3A and C_4AF can also bind the chloride ions and transfer into Friedel's salt. Usually, binding occurs instantaneously or at a much greater rate than transport velocities. The pore system is always considered to be at equilibrium. This assumption may be valid, when the chloride ion travels slowly, just in the case of diffusion alone, but this may not be valid when the ions are moving quickly and the test duration is short, as in the case of the rapid migration test. In this case, transport would be occurring too quickly for equilibrium to be maintained (Barbarulo et al. 2000; Samson et al. 2003). Tang and Nilsson (1993) reported that when crushed particles (with the size of 0.25–0.2 mm) were immersed in chloride solution, chemical binding almost completed after less than 14 days. However, Arya et al. (1990) found that bound chlorides were still increasing after 84 days of immersion in 2% chloride solution. Olivier (2000) believed that the rate of chloride binding on crushed mortar particles was very high. Indeed, more than 80% of the bound chlorides are bound in less than 5 h.

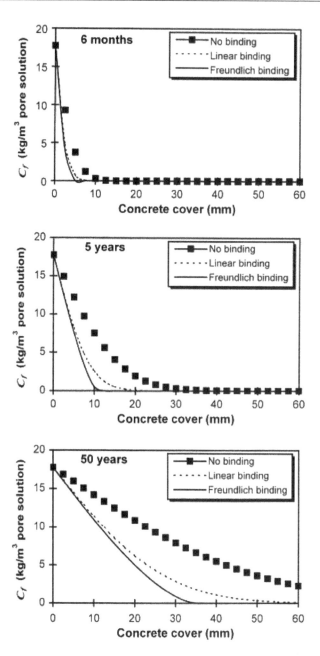

Figure 1.4 Free chloride concentration profiles at (a) 6 months, (b) 5 years, and (c) 50 years for 0.5 M exposure conditions. (From Martın-Pérez et al. 2000.)

The research on chloride binding has been carried out for a long time and in different cementitious systems, including cement-based materials (Gbozee et al. 2018; Shi et al. 2016) or alkali-activated materials (Ke et al. 2017a, b). In 1998, Justnes (1998) did an excellent work on reviewing the chloride binding in cementitious materials. The effects of cement type, mineral additives or replacement, cement content, water to binder ratio, curing and exposure condition, and chloride source on chloride binding of cementitious materials were discussed in detail. Besides, many experimental and review papers have also studied on different factors (De Weerdt et al. 2014, 2015; Florea and Brouwers 2012) which can affect the chloride binding capacity of cementitious materials. According to these studies, it has been confirmed that the content of C_3A and C_4AF dominate the chemical binding of chloride ion, while C_3S and C_2S dominate physical adsorption. Hydroxyl and sulfate ions may decrease the chloride binding capacity of cementitious materials.

1.4 ORGANIZATION OF THIS BOOK

This book summarizes recent progress in chloride-related issues in cement-based materials and structures. Focused on chloride-related corrosion occurring in cement concrete structures, the chloride ingress process, interaction between chloride ions and cement hydrates, influential factors, and testing methods for chloride transport in cement concrete are presented in detail. A brief introduction on the background of the topic of every chapter is firstly presented and is followed by a detailed summary of all related aspects of the topic.

1.4.1 Chapter 2—Mechanisms of Chloride Transport in Cement-Based Materials

In this chapter, the chloride transport process in cement-based materials is studied, and different mechanisms based on the driven forces are presented. The mechanism and influential factors of different chloride transport methods, including hydrostatic advection, capillary suction, diffusion, thermal and electrical migration, are discussed. The contents of this chapter can help to understand the transport process of chloride ions in cement-based materials and also lay the foundation for discussions in the chapters that follow it.

1.4.2 Chapter 3—Chemical and Physical Interactions between Chlorides and Cement Hydrates

In this chapter, mechanisms of chloride binding, including chemical binding and physical adsorption, are presented. Formation and stability of Friedel's salt are introduced as the main factors controlling the chemical-binding capacity of cement-based materials. The properties of electrical double layer formed at solid–liquid interface are discussed to explain the phenomenon of "chloride concentrate," which can be considered as an instable physical adsorption between hydration products and chloride ions.

1.4.3 Chapter 4—Chloride Binding and Its Effects on Characteristics of Cement-Based Materials

In this chapter, different forms of binding isotherm used to describe the chloride binding capacity of cement-based materials are presented. The application of binding isotherms in chloride binding capacity evaluation is discussed. The effects of chloride binding on microstructure and properties of cement concrete structure are summarized.

1.4.4 Chapter 5—Testing Methods for Chloride Transport in Cement-Based Materials

In this chapter, classifications of testing method for chloride transport in cement-based materials are reported. Testing procedure, parameters, and influential factors of these measurements are overviewed. Relationship and comparison of testing results of these different methods are also provided.

1.4.5 Chapter 6—Determination of Chloride Penetration in Cement-Based Materials Using $AgNO_3$-Based Colorimetric Methods

As an important testing method on chloride penetration depth determination, $AgNO_3$ colorimetric method has been widely applied. The factors affecting the testing results for penetration depth and chloride concentration at the color change boundary are discussed. The applications of $AgNO_3$ colorimetric method in chloride diffusion and migration coefficients determination are also reviewed.

1.4.6 Chapter 7—Factors Affecting Chlorides Transport in Cement-Based Materials

Chloride transport is a complex process in which chemical and physical reactions simultaneously occur. During the chloride transport process, any changes of properties, microstructure of materials, and environmental conditions can cause effects on the chloride transport process. In this chapter, factors that may influence the chloride transport of cement-based materials, including ion interaction, microstructure, chloride binding, and cracking are introduced and discussed in detail.

1.4.7 Chapter 8—Simulation and Modeling of Chloride Transport in Cement-Based Materials

Chloride transport is one of the main aspects in the establishment of service life prediction of concrete structures, and different models simulating the chloride transport process in cement-based materials have been proposed. Based on the knowledge and discussion in the previous chapters about chloride transport, chloride binding, and their relationship with the microstructure of the structure, this chapter summarizes different chloride transport models proposed for cement-based materials and their application in different materials and environmental conditions.

REFERENCES

Alexander MG, Nganga G. Reinforced concrete durability: Some recent developments in performance-based approaches. *Journal of Sustainable Cement-Based Materials*. 2014, 3(1):1–12.

Angst UM, Elsener B, Larsen CK, Vennesland Ø. Chloride induced reinforcement corrosion: Electrochemical monitoring of initiation stage and chloride threshold values. *Corrosion Science*. 2011, 53:1451–1464.

Ann K, Ahn J, Ryou J. The importance of chloride content at the concrete surface in assessing the time to corrosion of steel in concrete structures. *Construction and Building Materials*. 2009, 23:239–245.

Ann KY, Song H-W. Chloride threshold level for corrosion of steel in concrete. *Corrosion Science*. 2007, 49:4113–4133.

Arya C, Buenfeld N, Newman J. Factors influencing chloride-binding in concrete. *Cement and Concrete Research*. 1990, 20:291–300.

Barbarulo R, Marchand J, Snyder KA, Prené S. Dimensional analysis of ionic transport problems in hydrated cement systems: Part 1. Theoretical considerations. *Cement and Concrete Research*. 2000, 30:1955–1960.

Bazant ZP. Physical model for steel corrosion in concrete sea structures-theory. *ASCE Journal of the Structural Division*. 1979, 105:1137–1153.

Berke N, Bentur A, Diamond S. *Steel Corrosion in Concrete: Fundamentals and Civil Engineering Practice*. CRC Press; 2014.

Bertocci U, Ye YX. An examination of current fluctuations during pit initiation in Fe-Cr alloys. *Journal of the Electrochemical Society*. 1984, 131:1011–1017.

Borade AN, Kondraivendhan B. Corrosion behavior of reinforced concrete blended with metakaolin and slag in chloride environment. *Journal of Sustainable Cement-Based Materials*. 2019:1–20.

Borg RP, Cuenca E, Gastaldo Brac EM, et al. Crack sealing capacity in chloride-rich environments of mortars containing different cement substitutes and crystalline admixtures. *Journal of Sustainable Cement-Based Materials*. 2018, 7(3):141–159.

Budelmann H, Holst A, Wichmann H-J. Non-destructive measurement toolkit for corrosion monitoring and fracture detection of bridge tendons. *Structure and Infrastructure Engineering*. 2014, 10:492–507.

Burstein G, Pistorius P, Mattin S. The nucleation and growth of corrosion pits on stainless steel. *Corrosion Science*. 1993, 35:57–62.

Castellote M, Andrade C, Alonso C. Chloride-binding isotherms in concrete submitted to non-steady-state migration experiments. *Cement and Concrete Research*. 1999, 29:1799–1806.

Castellote M, Andrade C, Alonso C. Measurement of the steady and non-steady-state chloride diffusion coefficients in a migration test by means of monitoring the conductivity in the anolyte chamber. Comparison with natural diffusion tests. *Cement and Concrete Research*. 2001, 31:1411–1420.

Chao C, Lin L, Macdonald D. A point defect model for anodic passive films I. Film growth kinetics. *Journal of the Electrochemical Society*. 1981, 128:1187–1194.

Chaudhari S, Patil P. Inhibition of nickel coated mild steel corrosion by electro-synthesized polyaniline coatings. *Electrochimica Acta*. 2011, 56:3049–3059.

Chen Y, Shui Z, Chen W, et al. Chloride binding of synthetic Ca–Al–NO3 LDHs in hardened cement paste. *Construction and Building Materials*. 2015, 93:1051–1058.

Cheng Y, Wilmott M, Luo J. The role of chloride ions in pitting of carbon steel studied by the statistical analysis of electrochemical noise. *Applied Surface Science*. 1999, 152:161–168.

Dawson J, Ferreira M. Electrochemical studies of the pitting of austenitic stainless steel. *Corrosion Science*. 1986, 26:1009–1026.

De Weerdt K, Colombo A, Coppola L, Justnes H, Geiker MR. Impact of the associated cation on chloride binding of Portland cement paste. *Cement and Concrete Research*. 2015, 68:196–202.

De Weerdt K, Orsáková D, Geiker M. The impact of sulphate and magnesium on chloride binding in Portland cement paste. *Cement and Concrete Research*. 2014, 65:30–40.

DeBerry DW. Modification of the electrochemical and corrosion behavior of stainless steels with an electroactive coating. *Journal of the Electrochemical Society*. 1985, 132:1022–1026.

Eichler T, Isecke B, Bäßler R. Investigations on the re-passivation of carbon steel in chloride containing concrete in consequence of cathodic polarisation. *Materials and Corrosion*. 2009, 60:119–129.

Ekolu S, Thomas M, Hooton R. Pessimum effect of externally applied chlorides on expansion due to delayed ettringite formation: Proposed mechanism. *Cement and Concrete Research*. 2006, 36:688–696.

Elakneswaran Y, Nawa T, Kurumisawa K. Electrokinetic potential of hydrated cement in relation to adsorption of chlorides. *Cement and Concrete Research*. 2009, 39:340–344.

Florea M, Brouwers H. Chloride binding related to hydration products: Part I: Ordinary Portland cement. *Cement and Concrete Research*. 2012, 42:282–290.

Galvele J, Lumsden J, Staehle R. Effect of molybdenum on the pitting potential of high purity 18% Cr ferritic stainless steels. *Journal of the Electrochemical Society*. 1978, 125:1204–1208.

Galvele JR. Transport processes and the mechanism of pitting of metals. *Journal of the Electrochemical Society*. 1976, 123:464–474.

Gbozee M, Zheng K, He F, Zeng X. The influence of aluminum from metakaolin on chemical binding of chloride ions in hydrated cement pastes. *Applied Clay Science*. 2018, 158:186–194.

Ghods P, Isgor OB, Bensebaa F, Kingston D. Angle-resolved XPS study of carbon steel passivity and chloride-induced depassivation in simulated concrete pore solution. *Corrosion Science*. 2012, 58:159–167.

Glass G, Buenfeld N. The presentation of the chloride threshold level for corrosion of steel in concrete. *Corrosion Science*. 1997, 39:1001–1013.

Glass G, Buenfeld N. The influence of chloride binding on the chloride induced corrosion risk in reinforced concrete. *Corrosion Science*. 2000, 42:329–344.

Hirao H, Yamada K, Takahashi H, Zibara H. Chloride binding of cement estimated by binding isotherms of hydrates. *Journal of Advanced Concrete Technology*. 2005, 3:77–84.

Hoar T. The production and breakdown of the passivity of metals. *Corrosion Science*. 1967, 7:341–355.

Hoar T, Jacob W. Breakdown of passivity of stainless steel by halide ions. *Nature*. 1967, 216:1299.

Hou CC, Han LH, Wang QL, et al. Flexural behavior of circular concrete filled steel tubes (CFST) under sustained load and chloride corrosion[J]. *Thin-Walled Structures*. 2016, 107:182–196.

Ipavec A, Vuk T, Gabrovšek R, Kaučič V. Chloride binding into hydrated blended cements: The influence of limestone and alkalinity. *Cement and Concrete Research*. 2013, 48:74–85.

Iqbal PON, Ishida T. Modeling of chloride transport coupled with enhanced moisture conductivity in concrete exposed to marine environment. *Cement and Concrete Research*. 2009, 39:329–339.

Ishida T, Iqbal PON, Anh HTL. Modeling of chloride diffusivity coupled with non-linear binding capacity in sound and cracked concrete. *Cement and Concrete Research*. 2009, 39:913–923.

Jafarzadeh S, Adhikari A, Sundall P-E, Pan J. Study of PANI-MeSA conducting polymer dispersed in UV-curing polyester acrylate on galvanized steel as corrosion protection coating. *Progress in Organic Coatings*. 2011, 70:108–115.

Jin W, Zhang Y, Lu Z. Mechanism and mathematic modeling of chloride permeation in concrete under unsaturated state. *Journal of the Chinese Ceramic Society*. 2008, 36:1362–1369.

Johansen HD, Brett CM, Motheo AJ. Corrosion protection of aluminium alloy by cerium conversion and conducting polymer duplex coatings. *Corrosion Science*. 2012, 63:342–350.

Justnes H. A review of chloride binding in cementitious systems. *Nordic Concrete Research Publications*. 1998, 21:48–63.

Karpakam V, Kamaraj K, Sathiyanarayanan S, Venkatachari G, Ramu S. Electrosynthesis of polyaniline–molybdate coating on steel and its corrosion protection performance. *Electrochimica Acta*. 2011, 56:2165–2173.

Ke X, Bernal SA, Hussein OH, Provis JL. Chloride binding and mobility in sodium carbonate-activated slag pastes and mortars. *Materials and Structures*. 2017a, 50:252.

Ke X, Bernal SA, Provis JL. Uptake of chloride and carbonate by Mg-Al and Ca-Al layered double hydroxides in simulated pore solutions of alkali-activated slag cement. *Cement and Concrete Research*. 2017b, 100:1–13.

Kraljić M, Mandić Z, Duić L. Inhibition of steel corrosion by polyaniline coatings. *Corrosion Science*. 2003, 45:181–198.

Krishnakumark BP. Evaluation of chloride penetration in OPC concrete by silver nitrate solution spray method. *International Journal of ChemTech Research*. 2014, 6:2676–2682.

Kuang D, Cheng Y. Understand the AC induced pitting corrosion on pipelines in both high pH and neutral pH carbonate/bicarbonate solutions. *Corrosion Science*. 2014, 85:304–310.

Li L, Easterbrook D, Xia J, et al. Numerical simulation of chloride penetration in concrete in rapid chloride migration tests. *Cement and Concrete Composites*. 2015, 63:113–121.

Liao X, Cao F, Zheng L, et al. Corrosion behaviour of copper under chloride-containing thin electrolyte layer. *Corrosion Science*. 2011, 53:3289–3298.

Lou X, Singh PM. Role of water, acetic acid and chloride on corrosion and pitting behaviour of carbon steel in fuel-grade ethanol. *Corrosion Science*. 2010, 52:2303–2315.

Martın-Pérez B, Zibara H, Hooton R, Thomas M. A study of the effect of chloride binding on service life predictions. *Cement and Concrete Research*. 2000, 30:1215–1223.

Okamoto G. Passive film of 18–8 stainless steel structure and its function. *Corrosion Science*. 1973, 13:471–489.

Olivier T. Prediction of chloride penetration into saturated concrete—Multi-species approach: PhD thesis, Department of Building Materials, Chalmers University of Technology, Goteborg, Sweden; 2000.

Otieno M, Beushausen H, Alexander M. Chloride-induced corrosion of steel in cracked concrete—Part I: Experimental studies under accelerated and natural marine environments. *Cement and Concrete Research*. 2016a, 79:373–385.

Otieno M, Beushausen H, Alexander M. Chloride-induced corrosion of steel in cracked concrete—Part II: Corrosion rate prediction models. *Cement and Concrete Research*. 2016b, 79:386–394.

Pack S-W, Jung M-S, Song H-W, Kim S-H, Ann KY. Prediction of time dependent chloride transport in concrete structures exposed to a marine environment. *Cement and Concrete Research*. 2010, 40:302–312.

Pour-Ali S, Dehghanian C, Kosari A. Corrosion protection of the reinforcing steels in chloride-laden concrete environment through epoxy/polyaniline–camphorsulfonate nanocomposite coating. *Corrosion Science*. 2015, 90:239–247.

Richards LA. Capillary conduction of liquids through porous mediums. *Physics*. 1931, 1:318–333.

Richardson J, Wood G. A study of the pitting corrosion of Al byscanning electron microscopy. *Corrosion Science*. 1970, 10:313–323.

Rosenfeld I, Marshakov I. Mechanism of crevice corrosion. *Corrosion*. 1964, 20:115t–125t.

Samson E, Marchand J, Snyder KA. Calculation of ionic diffusion coefficients on the basis of migration test results. *Materials and Structures*. 2003, 36:156–165.

Samson E, Marchand J, Snyder KA, Beaudoin J. Modeling ion and fluid transport in unsaturated cement systems in isothermal conditions. *Cement and Concrete Research*. 2005, 35:141–153.

Sathiyanarayanan S, Azim SS, Venkatachari G. Corrosion protection coating containing polyaniline glass flake composite for steel. *Electrochimica Acta.* 2008, 53:2087–2094.

Sato N. A theory for breakdown of anodic oxide films on metals. *Electrochimica Acta.* 1971, 16:1683–1692.

Sergi G, Yu S, Page C. Diffusion of chloride and hydroxyl ions in cementitious materials exposed to a saline. *Magazine of Concrete Research.* 1992, 44:63–69.

Shan C, Hou X, Choy K-L. Corrosion resistance of TiO2 films grown on stainless steel by atomic layer deposition. *Surface and Coatings Technology.* 2008, 202:2399–2402.

Shi C, Hu X, Wang X, Wu Z, Schutter Gd. Effects of chloride ion binding on microstructure of cement pastes. *Journal of Materials in Civil Engineering.* 2016, 29:04016183.

Shi Z, Geiker MR, De Weerdt K, et al. Role of calcium on chloride binding in hydrated Portland cement–metakaolin–limestone blends. *Cement and Concrete Research.* 2017, 95:205–216.

Singh JK, Singh DDN. The nature of rusts and corrosion characteristics of low alloy and plain carbon steels in three kinds of concrete pore solution with salinity and different pH. *Corrosion Science.* 2012, 56:129–142.

Song H-W, Lee C-H, Ann KY. Factors influencing chloride transport in concrete structures exposed to marine environments. *Cement and Concrete Composites.* 2008, 30:113–121.

Spiesz P, Brouwers H. Influence of the applied voltage on the Rapid Chloride Migration (RCM) test. *Cement and Concrete Research.* 2012, 42:1072–1082.

Spiesz P, Brouwers H. The apparent and effective chloride migration coefficients obtained in migration tests. *Cement and concrete Research.* 2013, 48:116–127.

Tang L, Nilsson L-O. Chloride binding capacity and binding isotherms of OPC pastes and mortars. *Cement and Concrete Research.* 1993, 23:247–253.

Tang Y, Zuo Y, Wang J, Zhao X, Niu B, Lin B. The metastable pitting potential and its relation to the pitting potential for four materials in chloride solutions. *Corrosion Science.* 2014, 80:111–119.

Tennakoon C, Shayan A, Sanjayan JG, et al. Chloride ingress and steel corrosion in geopolymer concrete based on long term tests. *Materials and Design.* 2017, 116:287–299.

Tittarelli F, Bellezze T. Investigation of the major reduction reaction occurring during the passivation of galvanized steel rebars. *Corrosion Science.* 2010, 52(3):978–983.

Toumi A, Francois R, Alvarado O. Experimental and numerical study of electrochemical chloride removal from brick and concrete specimens. *Cement and Concrete Research.* 2007, 37:54–62.

Urquidi M, Macdonald DD. Solute–vacancy interaction model and the effect of minor alloying elements on the initiation of pitting corrosion. *Journal of the Electrochemical Society.* 1985, 132:555–558.

Van Belleghem B, Kessler S, Van den Heede P, et al. Chloride induced reinforcement corrosion behavior in self-healing concrete with encapsulated polyurethane. *Cement and Concrete Research.* 2018, 113:130–139.

Voinitchi Da, Julien S, Lorente S. The relation between electrokinetics and chloride transport through cement-based materials. *Cement and Concrete Composites.* 2008, 30:157–166.

Waseda Y, Suzuki S, Waseda Y. *Characterization of Corrosion Products on Steel Surfaces.* Springer; 2006.

Xu W, Liu J, Zhu H. Pitting corrosion of friction stir welded aluminum alloy thick plate in alkaline chloride solution. *Electrochimica Acta.* 2010, 55:2918–2923.

Yang Z, Weiss WJ, Olek J. Water transport in concrete damaged by tensile loading and freeze–thaw cycling. *Journal of Materials in Civil Engineering.* 2006, 18:424–434.

Ye H, Jin N, Jin X, et al. Chloride ingress profiles and binding capacity of mortar in cyclic drying-wetting salt fog environments. *Construction and Building Materials.* 2016, 127:733–742.

Yuan Q, Shi C, De Schutter G, Audenaert K, Deng D. Chloride binding of cement-based materials subjected to external chloride environment—A review. *Construction and Building Materials.* 2009, 23:1–13.

Chapter 2

Mechanisms of Chloride Transport in Cement-Based Materials

2.1 INTRODUCTION

Chloride transport is a kind of mass transport process which has long history dating back to the 19th century (Middleman 1997; Basmadjian 2005). Mass transport may happen in many scenarios, such as diffusion, distillation, drying, and leaching. Thus, mass transport has long been a fundamental scientific issue for many disciplines. A great amount of knowledge has been generated on the subject of mass transport. Therefore, the knowledge of chloride transport can be developed on quite solid foundations. Since the 1980s, the durability issue of concrete infrastructures subjected to chloride environment has become a serious problem for infrastructures, because a large number of marine infrastructures were built in the 1940s. After tens of years in service, many concrete structures have shown signs of deterioration, and some of the structures have to be demolished. Many research groups around the world have been, and still are being, carried out on this issue. The understanding of chloride transport in concrete has been advanced progressively.

Chloride ion can be introduced into concrete in two ways (Yuan 2009; Trejo 2018): (1) coming from the ingredients of concrete (internal chloride); (2) penetration from external environment (external chloride), mainly from seawater and de-icing salt. The internal chloride can be well controlled by limiting chloride content of raw materials for concrete. However, the chloride content in the environment can't be controlled. The service life of reinforced concrete subjected to chloride environments mainly depends on the ingress process of chloride ions into concrete, and the subsequent steel corrosion induced by the ingress chloride. Thus, the transport of chloride in concrete has attracted a lot of attention.

Although many studies have been carried out on chloride-related topics, the understanding of chloride transport is still very challenging. As stated in Tang and Nilsson (2012), due to the complicated physical and chemical process, there are at least three aspects contributing to the complexity.

- The exposure conditions vary constantly. Take the concrete structure in marine environments, for example; although the chloride concentration in seawater is relatively constant from sea to sea, concrete structures in different zones are under completely different attack, such as fully submerged zone, tidal zone, splash zone, and marine fog zone. The chloride content at different zones is different, and the water saturation states are also different. Consequently, the deterioration severity of concrete at different sections is different.
- The structure and compositions of concrete are complicated and evolve with time. Modern concretes have complicated ingredients, including cement, mineral admixtures, chemical admixtures, sand, and aggregates. Raw materials for concrete cover a very wide range of materials; thus, the interactions among materials are quite complicated, and the quality of raw materials may also fluctuate to some extent.
- In real engineering, chloride ions may transport in different mechanisms, but at a very slow speed. It involves very complicated physical and chemical interactions. However, the transport mechanisms are not well understood, and the testing methods for chloride transport may also not be able to reflect the real transport process of chloride ions in practice, because changes in temperature, rain, and sunshine introduce very complicated coupled effects which cannot be reproduced in lab.

This chapter gives a detailed review on the basic transport mechanisms of chloride in cement-based materials.

2.2 TRANSPORT OF CHLORIDE IN CEMENT-BASED MATERIALS

From the viewpoint of physics, the transport of physical unit can roughly be classified into two categories:

(1) Gradient-driven transport

The physical laws that govern the transport of mass, energy, and electricity are based on the notion that the flow of these entities is induced by a driving potential. The type of gradients can be in different forms. Generally, it can be expressed in two ways (Basmadjian 2005). Firstly, the gradient changes along with the direction of flow, such as diffusion or conduction. Take diffusion, for example; ions or molecular diffuse into the matrix because of the concentration difference. The concentration difference varies along the depth, and thus the driving force varies along the depth. The common mass transports at a constant driving force in different disciplines are given in Table 2.1.

Table 2.1 Mass Transfer at Varied Driving Forces

Name	Process	Flux	Gradient
Fick's law	Diffusion	$N/A = -D\dfrac{dC}{dx}$	Concentration
Fourier's law	Conduction	$q/A = -k\dfrac{dT}{dx}$	Temperature
Alternative formulation		$q/A = -\dfrac{d(\rho C_p T)}{dx}$	Energy Concentration
Newton's Viscosity law	Molecular momentum transport	$F_x/A = \tau_{qx} = -\mu\dfrac{dv_x}{dy}$	Velocity
Alternative formulation		$F_x/A = \tau_{yx} = -v\dfrac{d(\rho v_x)}{dy}$	Momentum Concentration
Poiseuille's law	Viscous flow in a circular pipe	$q/A = v_x = -\dfrac{d^2}{32\,\mu}\dfrac{dp}{dx}$	Pressure
D'Arcy's law	Viscous flow in A porous medium	$q/A = v_x = -\dfrac{K}{\mu}\dfrac{dp}{dx}$	Pressure

Where A is area, m^2; N is molar flow rate, mol/s; C is concentration, mol/l; x is depth, m; q is heat flow, J/s; k is thermal conductivity, J/m s K; T is temperature, K; α is thermal diffusivity, m^2/s; ρ is density, kg/m³; C_p is heat capacity at constant pressure, J/kg K or J/mol K; τ is shear stress, Pa; Fx is force, N; μ is viscosity, Pa s; v is kinematic viscosity, m^2/s; v is velocity, m/s; K is permeability, m/s or m^2; P is pressure, Pa.

In another scenario, the driving gradient is fixed along the flow direction. Thus, the driving force can be obtained by dividing the potential over the distance, such as electrical field. When applying an electrical field to a sample, the electrical field is constant along the depth of the sample. The common mass transport at a fixed driving force in different disciplines are given in Table 2.2.

(2) Non-gradient-driven transport

Species are most often driven by gradient. Capillary effect is probably the only mechanism that is a non-gradient-driven transport. It is also called capillary motion, capillary action, or wicking. It is the ability of a liquid to flow in narrow pores without the assistance of, or even in opposition to, external forces like gravity. This phenomenon happens very often in daily life. It occurs because of intermolecular forces between the liquid and surrounding solid surfaces. If the diameter of the tube is sufficiently small, then the combination of surface tension and adhesive forces between the liquid and narrow pore wall act to propel the liquid, which then transports into the narrow pore.

Table 2.2 Mass Transfer at Fixed Driving Forces

Process	Flux or Flow	Driving Force	Resistance
Electrical current flow (Ohm's law)	$i = \Delta V / R$	ΔV	R
Convective mass transfer	$N / A = k_c \Delta C$	ΔC	$1 / k_c$
Convective heat transfer	$q / A = h \Delta T$	ΔT	$1 / h$
Flow of water due to osmotic pressure	$N_A / A = P_\omega \Delta \pi$	$\Delta \pi$	$1 / P_\omega$

Where i is electrical current, A; ΔV is voltage difference, v; k_c, mass transfer coefficient; ΔC is concentration difference; h is heat transfer coefficient, J/m^2 K; P_ω water permeability, mol/m^2 s; N molar flow rate, mol/s; $\Delta \pi$ osmotic pressure, Pa.

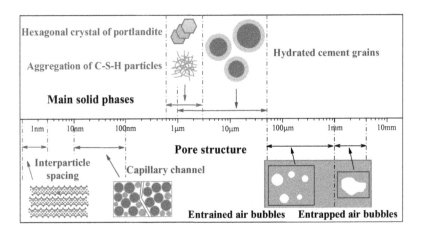

Figure 2.1 Dimensional range of pores in hardened cement paste.

Hardened cement paste and concrete are very porous materials. In addition to solids, the hydrated cement paste contains various types of pores. The porosity of cement paste may be up to 30–40% in volume, and the size of the pore varies from nm to mm. The porosity and pore structure have an important influence on the permeability of cement-based materials. The pores include interlayer space in calcium silicate hydrate (C-S-H), capillary voids, and air void (Figure 2.1) (Metha and Monteiro 2006).

- Interlayer space in C-S-H is assumed to be with the width of 5-25 Å and accounts for 28 percent porosity in solid C-S-H. This pore size is too small to have an adverse effect on the permeability of the hydrated cement paste.
- Capillary voids are the space not filled by the solid phases of the hydrated cement paste, ranging from 10 nm to several µm. It is

generally agreed that capillary pores smaller than 50 nm have no effect on the permeability of hardened cement paste; only the capillary pores larger than 50 nm have the effects.

- Air voids include entrained or entrapped air. Entrained air voids usually range from 50 to 200 μm. Entrapped air voids may be as large as 3 mm. Both of them have effect on permeability of hardened cement paste.

It is generally believed that there are two kinds of pores in cement-based materials, i.e., open (connected) pores and closed (dis-connected) pores. Chloride ions from the external environment can penetrate into cement-based materials through connected pores. In different circumstances, chloride can be transported into cement-based materials by different mechanisms. Chloride penetration is closely related to water permeability of cement paste. However, chloride ion is a charged particle, which is different from water molecular. As an ion, the movement of chloride ions can be driven by many forces. Some driven forces occur in real engineering, while others are only special techniques which are used to accelerate the transport of chloride ions.

As stated above, the transport of chloride ions in hardened cement-based materials can occur through many mechanisms (Yuan 2009; Fenaux 2019), such as diffusion, electrical migration hydrostatic advection or permeation, capillary effect, and thermal migration. In many real engineering cases, chloride ions are transported into concrete by coupled effects of several mechanisms. The following sections describe these transport mechanisms in detail.

2.2.1 Electrical Migration

When an external electrical field exists, chloride ions in the solution will move rapidly towards the positive electrode. This principle has been widely applied in accelerated test methods for the migration of chloride ions in concrete. This will be discussed in detail later. Also, it has been used to remove chloride ions in chloride ion–contaminated concrete (Orellan et al. 2004; Toumi et al. 2007). When ions move under the action of applied voltage, it also results in a difference in the concentration of these ions. This means that diffusion will also happen. Mathematically, the migration of a specific ion can be expressed as follows (Andrade 1993):

$$J(x) = -D\left[\frac{\partial c}{\partial x} + \frac{ZF}{RT} \cdot c \cdot \frac{\partial E}{\partial x}\right] \tag{2.1}$$

where D is the diffusion coefficient, the first term in the right-hand side is the diffusion term. It is often neglected when the external electrical field is high enough.

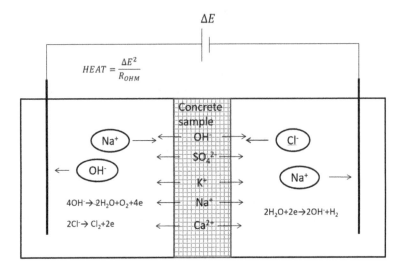

Figure 2.2 Illustration of migration processes and reactions happening in an electrically accelerated chloride migration test.

Figure 2.2 illustrates a testing setup for accelerated chloride migration testing cell under an applied field, which consists of two chamber cells and a concrete sample as the division between the two chambers. Due to the application of an electrical field, some electrolysis reactions also happen in the two chamber cells. It is worth mentioning that there is no electrolysis of water in positive electrodes, which, however, appeared on Andrade's (1993) figure. If the electrodes are inert materials, then in the positive chamber, the following reactions will occur:

$$4OH^- \rightarrow 2H_2O + O_2 + 4e \qquad (2.2)$$

$$2Cl^- \rightarrow Cl_2 + 2e \qquad (2.3)$$

If iron is used as positive electrode, iron may release electrons due to electrochemical reactions. This may result in corrosion on the positive electrode.

$$Fe \rightarrow Fe^{3+} + 3e \qquad (2.4)$$

In the negative chamber cell, the following reaction occurs:

$$2H_2O + 2e \rightarrow 2OH^- + H_2 \qquad (2.5)$$

If the applied voltage is high, it may generate a significant amount of heat. This may cause a change in microstructure of the concrete specimen.

Under the action of electrical field, chloride ions may transport in concrete in another scenario. To remove the chloride in concrete from the contaminated part, an electrical field can be applied and the chloride in concrete can be driven out of the concrete, i.e., electrochemical chloride extraction method (ECE) (Polder 1994; Siegwart et al. 2003; Elsener and Angst 2007; Zheng et al. 2016; Luan et al. 2017). As a consequence, the service life of contaminated structures can be extended. This method can remove chloride from concrete without destroying concrete, and thus has little effect on the environment and high efficiency. The ECE method consists of applying an electrical field between the reinforced bars inside the concrete structure and an externally placed electrode constituted by a metal mesh. Under the action of electrical field, charged ions such as chloride ions are attracted to the anode placed on the external surface of the concrete. Also, cations (i.e., Ca^{2+}, K^+, and Na^+) migrate into the concrete and hydroxyl ions (OH^-) may also form on the surface of reinforced bar. Figure 2.3 shows a schematic diagram of the chloride extraction technique.

In ECE technique, different electrolysis methods take place in anode and cathode. The anodic reaction is:

$$2OH^- = \frac{1}{2}o_2 + H_2O + 2e^-$$

$$2H_2O = o_2 + 4H^+ 4e^- \qquad (2.6)$$

$$2Cl^- = Cl_2 + 2e^-$$

Hydroxide ions are formed at the cathode. This further decreases the corrosion risk of steel.

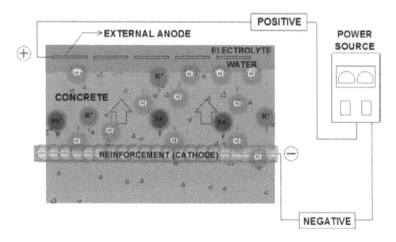

Figure 2.3 Schematic design of ECE method. (From Luan et al. 2017.)

$$\frac{1}{2}O_2 + H_2O + 2e^- = 2OH^-$$

(2.7)

$$2H_2O + 2e^- = H_2 + 2OH^-$$

The solutions most often used as external electrolytes are saturated calcium hydroxide, sodium hydroxide aqueous solutions. The current density used in ECE technique most ranges from 1 A/m² to 5 A/m² (Siegwart et al. 2003; Elsener and Angst 2007; Luan et al. 2017).

According to Orellan et al. (2004), chloride content after treatment for one week was reduced by around 40%, meanwhile, significant amounts of alkali ions were observed around the steel. Efficiency percentages can reach up to 70% according to experimental works.

2.2.2 Diffusion

When concrete is fully saturated, the transport of chloride in concrete is commonly assumed to be governed by diffusion. Chloride ions are diffusing in a porous matrix, instead of a homogeneous solution, that has both solid and liquid components. Chloride ions travel in the continuous liquid phase under concentration gradient in a "random walk" manner, as shown in Figure 2.4. The diffusion through the solid portion of the matrix is negligible when compared to the rate of diffusion through the pore solution. When the transport path is blocked by solid phase, chloride will not go any further and transport around the solid phase. The rate of diffusion is thus controlled not only by the diffusion coefficient through the pore solution but also by the physical characteristics of the capillary pore structure.

In the case of diffusion, the electrochemical gradient is the only driving force. The electrochemical potential of chloride is given by (Rieger 1994; Zhang and Gjorv 1994; Tang 1999; Poulsen and Mejlbro 2006; Fenaux 2019):

$$\mu = \mu_0 + RT \ln(\gamma c)$$

(2.8)

where μ is the electrochemical potential, μ_0 is the standard chemical potential, R is the universal gas constant, T is the absolute temperature, γ is the activity coefficient, c is the ionic concentration. The movement of ionic species is under the gradient of electrochemical potential, which is the combination of driving force (chemical potential) and draw-back force (counter- electrical field) (Tang 1999). It can be written as:

$$J = -\frac{D}{RT}c\nabla\mu = -D\frac{\partial c}{\partial x}\left(1 + \frac{\partial \ln \gamma}{\partial \ln c}\right) - cD\frac{zF}{RT}\frac{\partial E}{\partial x}$$

(2.9)

where J is the flux of ions, $\Delta\mu$ represents the gradient of electrochemical potential, D is the diffusion coefficient, z is the valence of ions, F is the

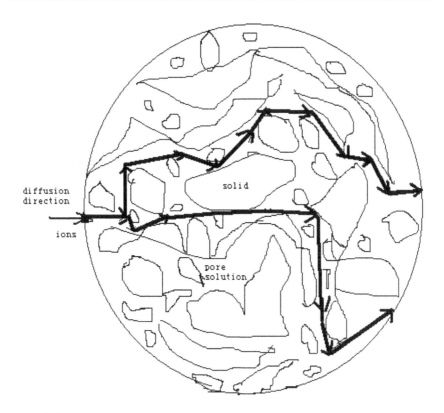

Figure 2.4 Illustration of ions travel through concrete at microscopic level.

Faraday constant, E is the electrical field, R is the gas constant, and x is the position variable.

For the purpose of simplification, in most literature, the terms $\dfrac{\partial \ln \gamma}{\partial \ln c}$ and $cD\dfrac{zF}{RT}\dfrac{\partial E}{\partial x}$ are often neglected. Equation 2.9 then becomes Fick's first law:

$$J = -D\frac{\partial c}{\partial x} \tag{2.10}$$

In practical terms, this equation is only useful after steady-state conditions have been reached, i.e., there is no change in concentration with time. It can be used, however, to derive the relevant equation for non-steady conditions (when concentrations are changing), often referred to as Fick's second law (Crank 1975):

$$\frac{\partial c}{\partial t} = \frac{\partial J}{\partial x} = -\frac{\partial c}{\partial x}\left(D \cdot \frac{\partial c}{\partial x}\right) = D \cdot \frac{\partial^2 c}{\partial x^2} \tag{2.11}$$

This equation has been solved using the boundary condition $c(x = 0, t > 0) = c_0$ (the surface concentration is constant at c_0), the initial condition $c(x > 0, t = 0) = 0$ (the initial concentration in the concrete is 0) and the infinite point condition $c(x = \infty, t > 0) = 0$ (far enough away from the surface, the concentration will always be 0). The classical error function solution is then obtained,

$$c(x,t) = c_0 \left[1 - \text{erf} \left(\frac{x}{\sqrt{4D}} \right) \right] \tag{2.12}$$

$$\text{erf}(z) = 2/\sqrt{\pi} \int_0^z \exp(-u^2) du \tag{2.13}$$

where erf is an error function. Equations (2.11) and (2.12) have been widely used to describe the diffusion of chloride ion in hardened concrete.

It can be seen that the terms of $\dfrac{\partial \ln \gamma}{\partial \ln c}$ and $cD\dfrac{zF}{RT}\dfrac{\partial E}{\partial x}$ in Equation 2.9 are assumed to be 0. This implicitly means that chloride ions are treated as "neutral particles," and travel in pore solution without the influence of other species. Obviously, this assumption is highly questionable. Some studies have shown that a change in the type of chloride solution, even though the chloride concentration was kept the same, may significantly influence the chloride penetration into concrete (Ushiyama and Goto 1974). This proves that the cation type of chloride salt influences the diffusion coefficient. Note that the pore solution of concrete is a concentrated solution with many types of ions, e.g., Na^+, K^+, SO_4^{2-}, OH^-, etc. which also have an influence on the chloride diffusion coefficient. Thus, to model chloride transport in concrete, chloride should be treated as negative particle instead of uncharged "neutral particle," and the interaction between the species has to be taken into account. Understanding the mechanism of ion diffusion in hardened cement paste is of great importance for predicting long-term durability of concrete structures. Gel pores in C-S-H forms dominant pathway for the transport of ions in cement paste with low water/cement (w/c) ratios where the electrical double layer effects play an important role (Yang et al. 2019). Experimental results suggest that the effective diffusivity of chloride ions is similar as that of tritiated water (HTO) and higher than the sodium ions. This difference can be attributed to the electrical double layer near the charged C-S-H surfaces. In order to understand the transport processes of different species in C-S-H and to quantify its effective diffusivity, a multiscale modeling technique has been proposed to combine atomic-scale and pore-scale modeling (Yang et al. 2019).

2.2.3 Hydrostatic Advection

For many infrastructures under seawater, the concretes are often saturated with water, and under a certain hydrostatic pressure. If there is a hydrostatic pressure exerted on the surface of the concrete and chloride ion is present in the solution, chloride ions will permeate into concrete along with the solution due to pressure gradients.

This phenomenon can be described by Darcy's law, which is an equation that describes the flow of a fluid through a porous medium, and initially was used for earth sciences. Darcy's law has the following form (Whitaker 1986):

$$Q = -\frac{\kappa A\left(p_b - p_a\right)}{\mu L} \tag{2.14}$$

where Q is the total discharge (m³/s), κ is the intrinsic permeability of the medium, A is the cross-sectional area to flow (m²), and the total pressure drop $(p_b - p_a)$ (pascals), μ is the viscosity of liquid (Pa·s) and L is the length over which the pressure drop is taking place (m).

A more general form of Darcy's law can be obtained:

$$q = -\frac{\kappa}{\mu}\nabla p \tag{2.15}$$

where q is the flow flux (m/s), ∇p is the pressure gradient vector (Pa/m).

Darcy's law is a simple and neat mathematical equation, and some basic principles can be deduced directly from the equations, as follows:

- If there is no pressure gradient over a distance, no flow occurs.
- If there is a pressure gradient, flow will occur from high pressure towards low pressure.
- The higher the pressure gradient across the porous materials, the greater the discharge rate.

2.2.4 Thermal Migration

As stated above, temperature gradient is also an important potential to drive the transport of mass. It is well known that ions or molecules in a hot environment move faster than those in a cold environment. There is one scenario in practice. If a saturated concrete sample has initial uniform chloride concentration all over the sample, when one portion is heated, chloride ion in the hotter region will similarly move towards colder regions. The most obvious situation when this process may occur is when a concrete structure, which has been contaminated with de-icing salt, heats up in sunlight. Salt-saturated water in the surface pores of the concrete will migrate rapidly into the cooler parts of the structure under the temperature gradient. This process is much more complicated, since it is the coupled effect of heat and mass. It should also be pointed out that the process is often combined with the process of diffusion.

2.2.5 Capillary Effect

Capillary effect is different from other mechanisms. The driving force for chloride ions is not potential but a free interfacial energy created by the difference between the forces that attract the molecules towards the interior of each phase and those that attract them to the contact surface. As the difference increases, fluids such as pore water are retained in the porous medium above the elevation of the water table. The interfacial tension or suction force in a capillary tube causes water to rise and form a meniscus. As stated above, the capillary effect can be described by the Laplace equation; the ingress of liquid depends on the surface tension of liquid, contact angle, and radius of pores. It is well known that the raw materials of concrete are hydrophilic, and the radius of capillary pores range from nm to mm. Thus, the solution containing chloride ions can ingress into concrete very quickly under the action of capillary effect. However, this transport mechanism is typically limited to a shallow cover region, and will not, by itself, bring chlorides to the level of the reinforcing steel unless the concrete is of extremely poor quality and the reinforcing steel is shallow. However, it does bring chlorides to some depth quickly in the concrete and reduces the distance that they must diffuse to reach the rebar (Thomas et al. 1995).

A common testing setup used to demonstrate the phenomenon is the capillary tube, as shown in Figure 2.5. When the lower end of a vertical glass tube is placed in water, a concave meniscus forms. This is because water cannot wet glass. However, if a concrete tube is placed in water, a convex meniscus forms. This is because water can wet concrete. When liquid can wet the capillary pore, the fluid and the capillary pore wall lift the liquid column up until there is a sufficient mass of liquid for gravitational forces to overcome this surface tension. The height that the liquid can lift up by

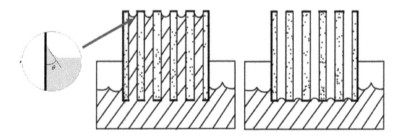

Figure 2.5 Capillary effect.

the surface tension depends on radius of the pore, the contact angle, and surface tension of liquid, and it can be described by Laplace's equation (Basford 2002).

$$h = \frac{2\gamma\cos\theta}{\rho gr} \text{ or}$$

$$P = \frac{2\gamma\cos\theta}{r}$$

(2.16)

2.2.6 Coupled Effects

In real engineering, the origins of chloride may be from seawater or de-icing salt. For concrete structures exposed to marine environments, the time of formwork removal may be much earlier than 28 days, for the sake of fast construction and economy. When concrete structures are first exposed to chloride environment, concrete is normally unsaturated with water and the hydration degree is relatively low. Thus, the exposure of concrete structures to marine environment at an early age make chloride ingress into concrete fast. In this scenario, the transport of chloride in concrete often involves several mechanisms, and the coupled mechanisms often interact with each other. For example, for unsaturated concrete structures, capillary effect first takes effect, and takes the solution containing chloride ions into the concrete cover. The depth where capillary effect can bring water in can be calculated by the Laplace equation. Or, capillary effect takes the solution into the depth where water is saturated. Afterwards, chloride ions penetrate concrete by diffusion alone.

On the other hand, concrete is a porous material, and chloride only transport in the pores which evolves with service time due to two reasons. First, continuous hydration of cement particles can decrease the porosity and densify the matrix, as shown in Figure 2.6. The porosity of concrete decreases with time, and the pore size also decreases with time. The porosity and pore size decrease fast at early age, and decrease slowly at later age. Second, concrete may suffer from different deteriorations during the period of service, and the pore structure may be coarsened by the deteriorations.

It is also worth it to mention that concrete structures along the seawater might be under the most complicated and severe chloride conditions. Based on the interaction between seawater and structure, or the distance between structures and seawater, the structures along the seawater may be classified into three zones, i.e., atmospheric zone, tidal zone, and immersed zone. For the atmospheric zone, the concrete structure interacts with fog containing chloride. Chloride ions precipitate on the concrete structure. In this case, concrete is often unsaturated. For the tidal zone, concrete is subjected to wetting–drying cycles. Under the action of wetting–drying cycles, chloride penetrates into concrete very fast. Concretes in the tidal zone are often the most deteriorated parts. In the case of the immersed zone, concrete is

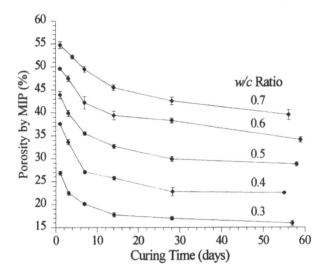

Figure 2.6 Effect of time and w/c ratio on total porosity determined by MIP. (From Metha et al. 2006.)

saturated with water, and chloride mainly transport into concrete by the mechanism of diffusion. It is worth it to mention that although the chloride concentration in concrete immersed in seawater is high, the corrosion of steel is rare, due to the lack of oxygen.

In cold climates, de-icing salts are often used to remove ice and snow. Reinforced concrete structures are often and severely damaged by the coupled effects of freezing–thawing cycles and de-icing salt attacks during winter periods. Freezing–thawing action can also cause additional water uptake known as frost suction. If a critical degree of water saturation is exceeded, severe deterioration of the microstructure of the concrete is likely, enhancing chloride ingress and increasing the probability of corrosion of its reinforcement (Kessler et al. 2017). It was found that the chloride penetration into concrete can be accelerated considerably by freezing–thawing cycles. The increasing of chloride diffusion coefficient can be described by a linear function of frost cycles. Under the combined actions of frost attack and chloride exposures, the propagation of steel corrosion was accelerated tremendously (Zhang et al. 2017). Obviously, the coupled effect of freezing–thawing makes the penetration of chloride into concrete more complicated.

2.3 SUMMARY

The transport of chloride in cement-based materials is a kind of mass transfer which has been an important topic for many disciplines, and a solid knowledge base has been generated in this field. As a charged particle, chloride may penetrate into hardened cement-based materials through many mechanisms, such as diffusion, electrical migration hydrostatic advection or permeation, capillary effect, and thermal migration. In real engineering, chlorides mainly come from seawater or de-icing salt. For different scenarios, various mechanisms may be involved in the penetration of chloride. In most cases, chloride ions are transported into concrete by coupled effects of several mechanisms. Due to the continuous evolution of cement-based materials and coupled transport mechanisms of chloride, the transport of chloride is a quite complicated process.

REFERENCES

Andrade C. Calculation of chloride diffusion coefficients in concrete from ionic migration measurements. *Cement and Concrete Research.* 1993, 23:724–742.

Basmadjian D. *Mass Transfer- Principles and Applications.* CRC Press, Boca Raton, FL; 2005.

Basford JR. The Law of Laplace and its relevance to contemporary medicine and rehabilitation. *Archives of Physical Medicine and Rehabilitation.* 2002,83:1165–1170.

Crank J. *The Mathematics of Diffusion,* 2nd ed. Oxford University Press, London; 1975.

Elsener B, Angst U. Mechanism of electrochemical chloride removal. *Corrosion Science.* 2007, 49:4504–2522.

Fenaux M, Reyes E, Gálvez JC, Moragues A, Modelling the transport of chloride and other ions in cement-based materials. *Cement and Concrete Composites.* 2019, 97:33–42.

Kessler S, Thiel C, Grosse CU, Gehlen C. Effect of freeze–thaw damage on chloride ingress into concrete. *Materials and Structures.* 2017, 50:121.

Lin H, Li Y., Li Y., A study on the deterioration of interfacial bonding properties of chloride-contaminated reinforced concrete after electrochemical chloride extraction treatment *Construction and Building Materials.* 2019, 197:228–240.

Luan R, Marcelo H, Eduardo P, Ana P. Electrochemical chloride extraction: Efficiency and impact on concrete containing 1% of NaCl. *Construction and Building Materials.* 2017, 145:435–444.

Metha PK, Monteiro PJM. *Concrete, Microstructure, Properties and Materials.* McGraw-Hill Press; 2006.

Middleman S, *Introduction to Mass and Heat Transfer.* John Wiley, New York; 1997.

Orellan JC, Escadeillas G, Arliguie G. Electrochemical chloride extraction: Efficiency and side effects. *Cement and Concrete Research.* 2004, 34:227–234.

Patel RA, Perko J, Jacques D, Schutter GD, Ye G, Bruegel KV, Effective diffusivity of cement pastes from virtual microstructures: Role of gel porosity and capillary pore percolation, *Construction and Building Materials.* 2018, 165:833–845.

Poulsen E, Mejlbro L. *Diffusion of Chloride in Concrete: Theory and Applications.* Taylor & Francis; 2006.

Polder RB. Electrochemical chloride removal of reinforced concrete prisms containing chloride from sea water exposure. UK Corrosion & Eurocorr'94, Boumnemouth, 1994.

Rieger PH. *Electrochemistry*, 2nd ed. Chapman & Hall, New York; 1994.

Siegwart M, Lyness JF, McFarland FJ. Change of pore size in concrete due to electrochemical chloride extraction and possible implications for the migration of ions. *Cement and Concrete Research*. 2003, 33:1211–1221.

Tang L. Concentration dependence of diffusion and migration of chloride ions Part 1. Theoretical considerations. *Cement and Concrete Research*. 1999, 29:1463–1468.

Tang L, Nilsson L, Muhammed Basheer PA. *Resistance of Concrete to Chloride Ingress: Testing and Modelling*. CRC Press; 2012.

Toumi A, François R, Alvarado O. Experimental and numerical study of electrochemical chloride removal from brick and concrete specimens. *Cement and Concrete Research*. 2007, 37:54–62.

Trejo D, Shakouri M, Pavan N, Vaddey O, Isgor B. Development of empirical models for chloride binding in cementitious systems containing admixed chlorides. *Construction and Building Materials*. 2018, 189:157–169.

Thomas MDA, Pantazopoulou SJ, Martin-Perez B. Service life modelling of reinforced concrete structures exposed to chlorides—A literature review, prepared for the Ministry of Transportation. University of Toronto, Toronto, ON, 1995.

Ushiyama H, and Goto S. Diffusion of various ions in hardened Portland cement pastes. 6th International Congress on the Chemistry of Cement, Moscow. 1974, 2:331–337.

Whitaker. S. Flow in porous media I: A theoretical derivation of Darcy's law. *Transport in Porous Media*. 1986, 1:3–25.

Yang Y, Patel RA., Churakov SV., Prasianakis NI, Kosakowski G, Wang M. Multiscale modeling of ion diffusion in cement paste: Electrical double layer effects. *Cement and Concrete Composites*, 2019, 96:55–65.

Yuan Q. Fundamental studies on test methods for the transport of chloride ions in cementitious materials. PhD thesis. Ghent University, Belgium, 2009.

Zhang P, Cong Y, Vogel M, et al. Steel reinforcement corrosion in concrete under combined actions: The role of freeze-thaw cycles, chloride ingress, and surface impregnation. *Construction and Building Materials*. 2017, 148:113–121.

Zhang T, Gjorv OE. An electrochemical method for accelerated testing of chloride diffusion. *Cement and Concrete Research*. 1994, 24:1534–1548.

Zheng L, Jones MR, Song Z. Concrete pore structure and performance changes due to the electrical chloride penetration and extraction. *Journal of Sustainable Cement-based Materials*. 2016, 5:76–90.

Chapter 3

Chemical and Physical Interactions between Chlorides and Cement Hydrates

3.1 INTRODUCTION

When chloride ions from environmental solutions penetrate into concrete, some of them are captured by the hydration products. This is called chloride binding. Chloride binding is significant to the study of the service life of reinforced concrete structures. Therefore, the effect of chloride binding must be taken into account when studying chloride ion transport in concrete. Concrete is a porous material that has both solid and liquid components. The mass transport through the solid portion of the matrix is negligible when compared to that through the pores. When concrete is saturated, the transport is controlled by the movement of mass through the pore solution. When concrete is not saturated, it is controlled by suction. In most cases, the service life of reinforced concrete depends on chloride ion penetrating from the external environment into the concrete.

Chlorides can exist either in the pore solution, chemically bound to hydration products, or physically held to the surface of hydration products. Some researchers believed that only free chloride ions in the pore solution are responsible for corrosion initiation. However, bound chloride may also be responsible for the corrosion initiation due to the release of some bound chloride ions into the pore solution under some specific conditions. Thus, total chloride, instead of free chloride, is sometimes used to evaluate corrosion risk of steel reinforcement in concrete. Due to the retardation effect of bound chloride, the free and bound chlorides must be distinguished from each other in service life prediction models. Chemical binding is generally the result of reaction between chlorides and C_3A or AFm phase to form Friedel's salt or the reaction with C_4AF to form a Friedel's salt analog. Physical binding is due to the adsorption of chloride ion to the calcium silicate hydrate (C-S-H) surfaces. Three types of interaction between chloride ions and C-S-H can be distinguished: in a chemisorbed layer on hydrated calcium silicates, present in the C-S-H interlayer spaces, and intimately bound in the C-S-H lattice. Generally, it is considered that the formation of Friedel's salt is the main mechanism of chloride binding and accounts most of the chloride binding. $Ca(OH)_2$ and Friedel's salt also played a role in chloride binding, even though only 2–5% in total. Within cement-based materials, chloride ions can be bound by different hydration products, including HO-AFm, SO_4-AFm, C-S-H gel, $Ca(OH)_2$, and Friedel's salt; even different chloride binding capacities of these hydrates are presented in various studies. Generally, the content of aluminum phase (C_3A and C_4FA) controls the chloride binding of cement-based materials internally mix with chloride salt, while when external chloride is applied, the chloride binding of cement-based materials is mostly due to the chemical binding of AFm phase and physical adsorption of C-S-H gel.

During the mass transport, chemically and physically bound chlorides are usually not distinguished from each other and chloride binding isotherm is generally applied to describe the chloride binding capacity of cement-based materials. In this chapter, the mechanisms of chemical binding and physical adsorption in cement-based materials are discussed. The formation of Friedel's salt and chloride concentrate phenomenon are reported in detail as the main mechanism of chemical binding and physical adsorption respectively.

3.2 THE FORMATION OF FRIEDEL'S SALT

Chemical binding of chloride ions generally relates to the chemical reactions between chloride ions in pore solution and hydrate cement or other solid phases in cement system. In cement system, chloride ions in pore solution have been reported to be captured by different hydrates or solid phases. As mentioned earlier, chloride ion can be introduced into concrete in two ways: (1) through ingredients of concrete (internal chloride); (2) penetration from external environment (external chloride). In most cases, the service life of reinforced concrete depends on chloride ion penetrating from the external environment into the concrete. External chloride affects chloride binding in a different way from internal chloride (Hassan 2001). Nagataki et al. (1993) found that the binding capacity of the intruded chloride is about two or three times as high as that of the internal chloride bound in the corresponding cementitious pastes.

Florea and Brouwers (2012) quantitatively studied the chloride binding capacity of hydration products of Portland cement in different external chloride concentration. The results indicated that HO-AFm phase made the greatest contribution for chloride binding under every external chloride ion concentration. However, the effects of HO-AFm phase on chloride binding of cement-based materials were decreased with the increase of external chloride concentration. All of the HO-AFm phases were transferred to Friedel's salt as the chloride concentration higher than 15 mmol/L. With the increase of chloride concentration, the role of SO_4-AFm phase in chloride binding became more and more important. According to the results of Florea and Brouwers (2012), the contribution of SO_4-AFm in chloride binding increased from 13% to 33% as the chloride concentration increased from 0.3 to 3.0 mol/L. The content of chloride ions bound by SO_4-AFm reached the same level of OH-AFm bound chloride when chloride concentration of external surrounding was high enough. Aluminum phase and AFm phase are two of the main phases to combine chloride ions and form Friedel's salt.

3.2.1 Reaction between Chloride Ions and Aluminum Phase

Much research has been done on chemical binding of chloride ions. Diamond et al. (Diamond 1986; Shi et al. 2017; Ke et al. 2017; Paul. et al. 2015) reported that C_3A can react with chloride ions to form Friedel's salt $(C_3A \cdot CaCl_2 \cdot 10H_2O)$. The research of Hewlett et al. (Hewlett 2003; Chen et al. 2015) revealed that the assumption that the content of aluminate was directly proportional to chemical binding capacity of cement only worked for internal chloride ions. Diamond et al. (Diamond 1986; Gong and Wang 2016; Suryavanshi and Swamy 1996; Yang et al. 2019) verified the formation of Friedel's salt both for internal and external chloride ions.

For internal chloride, it is widely believed that the C_3A phase in cement can react with chloride ions and form Friedel's salt during the process of cement hydration (Diamond 1986; Kim et al. 2016). However, the sulfates can compete with chloride ions to be captured by C_3A. Actually, when chloride salts are internally contained before cement hydration, sulfates have the priority to react with C_3A and form ettringite crystals $(Ca_6Al_2(SO_4)_3(OH)_{12} \cdot 26H_2O$, AFt) until the depletion of sulfates. After that, Friedel's salt can be formed with the reaction between chloride ions and C_3A. If the chloride ions are limited in the system, the AFt phase can transfer into monosulfoaluminate hydrate $(3CaO \cdot Al_2O_3 \cdot CaSO_4 \cdot nH_2O$, AFm) by reacting with the residual C_3A and C_4AF (Spice 2016; Glasser 2001; Midgley and Illston 1984). The study of Ekolu et al. (2006) revealed the mechanism of chloride binding by ettringite. Ettringite can be formed with the desorbed sulfate from C-S-H gel without chloride source. With the increase of chloride concentration, monosulfate will release sulfate ions with the attack of chlorides and accelerate the formation of ettringite between sulfate ions and unattacked monosulfate. However, when chloride concentration exceeds a threshold value, both monosulfate and ettringite will be destroyed by chlorides and transform into Friedel's salt and gypsum.

When external chlorides are applied, the chloride binding is different from internal chlorides as most of C_3A has been hydrated. Mehta et al. (1977) found that AFm and AFt wouldn't take part in the chloride binding for external penetrated chloride. Midgley and Illston (1984) reported that only the unhydrated C_3A could bind chloride ions and form Friedel's salt; different conclusions were given by Glasser (2001) that even hydrated C_3A could also react with chloride ions into Friedel's salt. Generally, the chemical binding capacity of cement-based materials is determined by the content of aluminate and aluminoferriate. According to the research of Wang et al. (2000), SO_4^{2-}, CO_3^{2-}, and Cl^- were generally coexisted within cement concrete, and the chloride ions could only react with the remained aluminate after reacting with sulfate and carbonate. The chloride binding capacity of cement-based materials related to the effective aluminate and aluminoferriate contents, which was the content of aluminate and aluminoferriate after consuming by sulfate and carbonate.

The reaction between internal $CaCl_2$ and C_3A in cement-based materials was proposed by Ben-Yair et al. (1974) with the chemical equation as follows:

$$C_3A + CaCl_2 + 10H_2O \rightarrow C_3A \cdot CaCl_2 \cdot 10H_2O \tag{3.1}$$

For internal NaCl salt, it will first react with $Ca(OH)_2$ as:

$$Ca(OH)_2 + 2NaCl = CaCl_2 + 2Na^+ + 2OH^- \tag{3.2}$$

After that, the chemical reaction as shown in Eq. (3.1) can occur. It was found that the addition of triisopropanolamine (TIPA) hastened the dissolution of aluminum into liquid phase and accelerated the formation of Friedel's salt (Ma et al. 2018).

The reaction between chloride ions and C_4AF was also studied by some researches (Csizmadia et al. 2001; Baroghel-Bouny et al. 2012; Xu et al. 2016), and the Friedel's analogue ($C_3F \cdot CaCl_2$) was considered to be formed. However, the significance of this reaction on chloride binding of cement-based materials is still unclear.

3.2.2 Reaction between Chloride Ions and AFm Phase

AFm is the catch-all item for compounds with same structure, which includes SO_4-AFm ($C_3A \cdot CaSO_4 \cdot 14H_2O$), HO-AFm ($C_3A \cdot Ca(OH)_2 \cdot 12H_2O$), CO_3-AFm ($C_3A \cdot CaCO_3 \cdot 10H_2O$), and the combination of any two of them. It is generally considered that AFm phases can chemically bind chloride (Ekolu et al. 2006; Glasser et al. 1999; Matschei et al. 2007). In hydrated cement-based materials, most of the AFm families always contain OH^-, SO_4^{2-} or CO_3^{2-} ions, which can be displaced by chloride ions (Balonis et al. 2010). With the substitution of chloride ions, AFm phases can convert to Friedel's salt or Kuzel's salt ($C_3A \cdot 1/2CaCl_2 \cdot CaSO_4 \cdot 10H_2O$). Ion exchange (Glasser et al. 1999; Suryavanshi et al. 1996), dissolution, and precipitation (Florea and Brouwers 2012) mechanisms have been applied in many researches to describe the transformation of AFm phase to Friedel's salt or Kuzel's salt. According to ion exchange mechanism, the OH^- located in the interlayer region of OH-AFm phase can be replaced by the internal or external chloride ions. Suryavanshi et al. (1996) described the chloride binding mechanism by the following equation:

$$R\text{-}OH^- + Na^+ + Cl^- \rightarrow R - Cl^- + Na^+ + OH^- \tag{3.3}$$

where R relates to the main compounds in the interlayer of OH-AFm phase. The molecular formula of R is $[Ca_2Al(OH^-)_6 \cdot nH_2O]^+$, the value of n is mostly controlled by the type of OH-AFm phase. Yonezawa (1989) explained the formation of Friedel's salt with internal chloride ions by ion exchange mechanism solely, while the study of Suryabanshi and Swamy (1996) found that the chloride binding by ion exchange only occupied a small fraction of total content of bound chloride. According to the definition of AFm phase, Friedel's salt is also a kind of AFm phase and can be noted as Cl-AFm. The lamellar molecular structure of $[Ca_2Al(OH^-)_6 \cdot nH_2O]^+$ can be understood based on the structure of $Ca(OH)_2$. Assume that 1 mol Ca^{2+} in 3 mol $Ca(OH)_2$ is replaced by 1 mol Al^{3+}, imbalance of charge occurs in the interlayer. Then, the chloride ions in pore solution can be absorbed and form Friedel's salt. When internal chloride salt is incorporated, Friedel's salt mainly precipitates during the cement hydration as a kind of AFm phase and only a small portion Friedel's salt may be formed by ion exchange from OH-AFm. Therefore, it can be concluded that the formation of Friedel's salt by chemical binding of internal chloride can be achieved by precipitation and ion exchange, and the former is the majority.

Balonis et al. (2010) introduced chloride ions into AFm phases with and without carbonate and studied the reaction products, as shown in Figure 3.1. They found that sulfate in SO_4-AFm can be replaced by chloride ions, and Kuzel's salt formed with low chloride concentration while Friedel's salt with high chloride concentration in solution. The displaced sulfate can transform into AFm phase by connecting with calcium and aluminum and lead to volumetric expansion and block of pores. Meanwhile, Friedel's salt can also be formed from the exchange of chloride ions and carbonate in CO_3-AFm phase. During the process of this replacement, no change will occur to molar volume, and the displaced carbonate can turn into calcite. Comparing the systems with and without carbonate, the carbonate may bring some effects on the activity of the introduced chloride ions. The carbonate can decrease the binding capacity of chloride ions and increase the value of $[Cl^-]/[OH^-]$ in the system.

Hirao et al. (2005) immersed synthetic AFm in different concentrations of NaCl solution and studied the content and composition of minerals by XRD test. As shown in Figure 3.2, a small diffraction peak was obtained at the left side of the expected diffraction peak of Friedel's salt. As the diffraction peaks relate to Friedel's salt and AFm phase is difficult to separate in XRD patterns, this study failed to determine whether the peak at the left

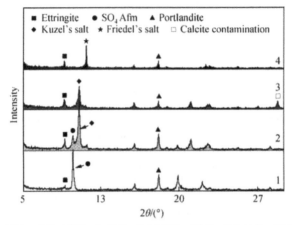

Legend: ■ Ettringite   ● SO₄Afm   ▲ Portlandite   ◆ Kuzel's salt   ★ Friedel's salt   □ Calcite contamination

1-Without CaCl₂; 2-0.003 mol of CaCl₂ added; 3-0.005 mol CaCl₂ added; 4-0.01 mol CaCl₂ added.

Figure 3.1 XRD patterns showing mineralogical changes and influence of CaCl₂ addition for 0.01 mole C₃A-0.0 mol CaSO₄-0.015 mol Ca(OH)₂-60 mL H₂O at (25 ±2) °C. (From Balonis et al. 2010.)

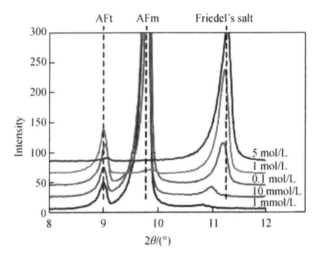

Figure 3.2 XRD patterns of AFm. (From Hirao et al. 2005.)

of Friedel's salt represented the Kuzel's salt. As the chloride concentration increased to 1 mol/L or greater, the AFm phase totally disappeared and a peak corresponding to Friedel's appeared. Considering the morphology properties of AFm phase before immersion and Friedel's salt after immersion in NaCl solution shown in Figure 3.3, the morphology of AFm phase was unchanged during chloride binding. It can be concluded that the process of transformation from AFm phase to Friedel's salt or chloride binding

Figure 3.3 Secondary electron images of AFm. (From Hirao et al. 2005.)

was mainly controlled by ion exchange mechanism under this condition. However, limited studies have been reported on formation of Kuzel's salt when chloride salt is internally incorporated. Chloride binding isotherms are mostly applied to represent the chloride binding capacity of AFm phase for ion exchange, precipitation, or other mechanisms.

According to the research of Elakneswaran et al. (2009a), Friedel's salt also has the ability to bind chloride ions. They found that the ionization of Friedel's salt gave rise to positive charge ($[Ca_2Al(OH)_6]^+$ mainly) of solid surface, which can be balanced by chloride ions in pore solution.

No matter whether chloride ions are internally or externally incorporated, the chemical binding in cement-based materials mainly relates to the formation of Friedel's salt. However, the mechanism of Friedel's salt formation is different for internal and external chlorides. For internal chloride,

chloride ions are bound with C₃A by dissolution and precipitation mechanism, while ion exchange between chloride ions and AFm controls the chloride binding when external chloride is added.

3.3 THE STABILITY OF FRIEDEL'S SALT

Yue et al. (2018) studied the characteristics of synthesized pure Friedel's salt by TG/DTG. As shown in Figure 3.4, a well-revolved TG curve with three weight-loss steps was obtained, which can be used to verify the composition of Friedel's salt (Birnin-Yauri and Glasser 1998; Vieille et al. 2003).

(1) Weight loss between 25 and 200°C was due to the removal of the water molecules in the interlayer space and a product with reduced crystallinity such as $3Ca(OH)_2.2Al(OH)_3$. $CaCl_2$ was formed in this process (Birnin-Yauri and Glasser 1998).

(2) Between 200 and 400°C, two weight loss peaks corresponding to the dehydroxylation of the main portlandite-like sheets were observed at about 260°C and 340°C respectively. During this process, a poorly structured phase was formed.

(3) With the temperature range from 400 to 1000°C, a slight weight loss was observed due to various phase transformations, such as the recrystallization of the amorphous phase (above 750°C) (Vieille et al. 2003), or the release of water from recombination of hydroxyl groups or probably anion decomposition.

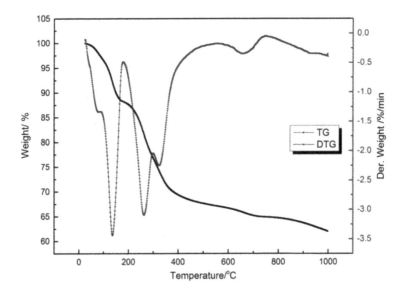

Figure 3.4 TG and DTG analysis of synthesised Friedel's salt. (From Yue et al. 2018.)

The stability of Friedel's salt is of great important for chloride binding of cement-based materials; the dissolution of Friedel's salt can release chloride ion into pore solution. Hassan (2001) investigated the stability of Friedel's salt and found that it was highly dependent on the chloride concentration in the surrounding solution. When chloride concentration decreased, Friedel's salt in the system could decompose into Kuzel's salt. It was also reported (Thomas et al. 2012) that Friedel's salt can be formed in cement paste when immersed in high concentration NaCl solution. As shown in Figure 3.5, as the chloride concentration decreased, the content of Friedel's salt detected by XRD gradually decreased and a new compound which may be Kuzel's salt was detected. Balonis et al. (2010) studied the effects of chloride ions on mineralogy of hydrated Portland cement systems, and proposed the schematic phase relation at 25°C between Friedel's salt and other AFm phases shown in Figure 3.6. It can be seen from the figure that most of composition ranges were dominated by two solid phases of SO_4-AFm and Friedel's salt. Kuzel's salt can be destabilised at very small amounts of carbonate and will only be encountered in low-carbonate environments.

Saikia et al. (2006) investigated the content of main hydration products C-S-H gel, calcium aluminum silicate hydrate (C-A-S-H) gel, and C_4AH_{13} in metakaolin(MK)-lime paste. When 2.5% of chloride ions were mixed in MK-lime paste, the main hydration products were C-S-H gel and Friedel's salt on the first 3d of hydration, while they gradually transferred into more stable C-A-S-H gel. Along with this process, the pH value of pore solution was unchanged. The re-decomposition of Friedel's salt was affected by formation of C-A-S-H gel during hydration process. When 5% and 10% of internal chlorides were added, the formation of Friedel's salt coincided with formation of a large number of C-S-H gels. For paste incorporated with 5% chloride ions, after 60d of hydration, a small number of C-A-S-H gels

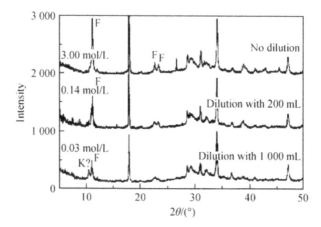

Figure 3.5 XRD patterns for pastes after immersion in 3 mol/L NaCl solution followed by immersion in different volumes of chloride-free solution. (From Thomas et al. 2012.)

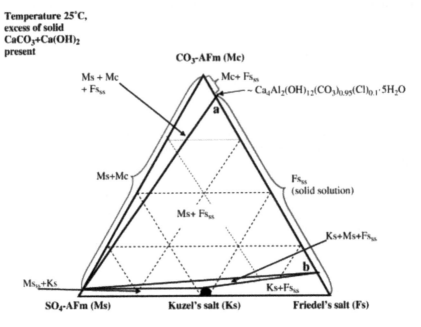

Figure 3.6 Schematic phase relations at 25°C, between Friedel's salt, monosulfoalumi-nate, and monocarboaluminate. (From Balonis et al. 2010.)

were still detected. Meanwhile, due to the formation of Friedel's salt, the hydration reaction of MK-lime paste was accelerated by internal chloride ions. At the later stages of hydration, no more chloride salt was generated from the dissolution of Friedel's salt. Therefore, the metal ions in the former chloride salts would react with some hydration products. When 2.5% chloride ions were internally mixed, the decrease of pH value (<12) was considered as the reason for dissolution of Friedel's salt and formation of more stable C-A-S-H. Suryavanshi and Swamy (1996) reported that the stability of Friedel's salt highly depended on pH value and chloride concentration of pore solution in cement-based materials. They explained this by the following equation:

$$3CaO \cdot Al_2O_3 \cdot CaCl_2 \cdot 10H_2O(s) \rightleftharpoons 4Ca^{2+}$$

$$+2Al(OH)^- + 2Cl^- + 4OH^- + 4H_2O$$

(3.4)

The carbonation reaction consumed a large amount of OH⁻ and resulted in the chemical reaction shifted to the right and increased the solubility of Friedel's salt. The decrease of chloride ions could also explained by this mechanism. When chloride concentration in cement pastes increases, the high stability of Friedel's salt can be also understand from Equation 3.4.

It can be known that the chloride concentration and pH value of the surrounding environment are two of main factors affecting the stability of Friedel's salt. The decreases of pH value and chloride concentration of pore solution will both result in the decomposition of Friedel's salt and may transfer to Kuzel's salt. However, the stability of Kuzel's salt is still not totally clear and further studies are needed.

3.4 PHYSICAL ADSORPTION OF CHLORIDE

3.4.1 Chloride Adsorption by C-S-H Gel

Diamond et al. (Diamond 1986; Ye et al. 2016; Shi et al. 2017) revealed that the chloride binding on the surface of C-S-H gels was the main part of physical adsorption. Hirao et al. (2005) studied the morphology properties of C-S-H gel before and after coming into contact with chloride ions. As shown in Figure 3.7, the morphology of C-S-H gels was almost the same with and without chloride adsorption; the physical adsorption of chloride didn't bring any changes on structural characteristics of C-S-H gel. Zibara et al. (2008) investigated the interaction between chloride ions and C-S-H gel in the system without aluminum phase. The results showed that the interaction between chloride ions and C-S-H gel was accelerated by higher w/b ratio of C-S-H gel. Even though the interaction between chloride ion and C-S-H gel was not that much greater than chloride binding by aluminum phase, the chloride adsorption by C-S-H gel was still important due to the relatively larger numbers of C-S-H gels compared to other hydration products. Tang and Nilsson (1993) found that most of the physical adsorbed chloride ions were adsorbed on the ion exchange sites of C-S-H gel and most of them were reversible. Therefore, the decrease of chloride concentration of pore solution can result into desorption of chloride ions from the exchange sites. On the surface of C-S-H gel, the adsorbed chloride ions can move from area to area by concentration gradient with a much lower speed to that in pore solution. Ramachandran (1971) divided the interaction between chloride ions and C-S-H gels into three types: in the chemical adsorption layer of C-S-H gel, penetrate in interlayer of C-S-H gel, and tightly bound in the lattice of C-S-H gel. Monteiro et al. (1997) investigated that the surface potential of C-S-H gel was generally determined by C/S ratio of C-S-H gel and positive potential was achieved by high C/S ratio. The anion ions in pore solution such as Cl- and OH- can be attracted and adsorbed on the surface of C-S-H gel with positive charged gel surface.

Nagataki et al. (Nagataki et al. 1993; Friedmann et al. 2012; Hu et al. 2018) explained the adsorption of C-S-H gel surface on chloride ions by the electrical double layer (EDL) theory. Laidler and Meiser (1982) stated that the surfaces of cement hydration products were negatively charged and

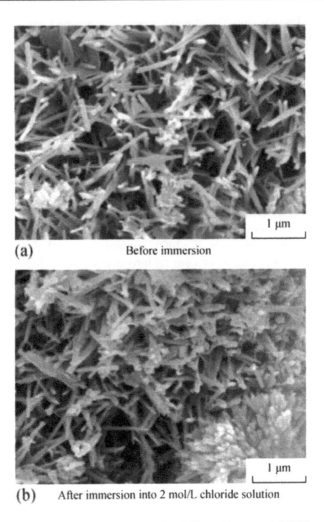

(a)   Before immersion

(b)   After immersion into 2 mol/L chloride solution

Figure 3.7 Secondary electron images of C–S–H. (From Hirao et al. 2005.)

could attract the cation ions such as Ca^{2+} and Na^{2+} in alkaline pore solution to form the so-called Stern layer or compact layer. The surface potential tended to be positive due to the Stern layer, and anion ions were attracted and diffusely distributed out of Stern layer to achieve potential neutrality in the area close to the solid surface or EDL. The capability of C-S-H gel for ion adsorption highly depends on the surface area of C-S-H gel and the potential differences between Stern and diffuse layers, which is also called zeta potential. The value of zeta potential is a function of valence of cation ions in Stern layer, temperature, and ion concentration of pore solution. Elakneswaran et al. (2009a) found that the value and sign of zeta potential of cement paste were mainly determined by Ca^{2+} concentration of pore solution. On the basis of Diamond et al. (1964), Larsen (1998) considered that the original zeta potential of C-S-H gel was positive; it became negative when swept away the $Ca(OH)_2$ in C-S-H until the pH value decreased to about 10. It can be concluded that the Ca^{2+} concentration of pore solution controls the zeta potential and C-S-H gel and thus affects the physical adsorption of chloride ions in cement-based materials. The formation and properties of EDL, and also the zeta potential of EDL for cement-based materials, will be detailed in the next section. Due to the relationship between physical adsorption and the surface potential of C-S-H gel, it can be imagined that the variations of interfacial properties between C-S-H gel and pore solution may have effects on physical adsorption. According to experimental results (Xu et al. 2016), the addition of different types of surfactants all changed the physical adsorption of cement hydrates, while few effects were obtained for chemical binding. The effects of surfactants on chloride adsorption depend on the surface potential of surfactants. It was found that the negative charged polycarboxylate superplasticizer decreased the chloride adsorption on C-S-H gel and a modified EDL model, considering the addition of polycarboxylate superplasticizer was proposed (Feng et al. 2018).

3.4.2 Chloride Adsorption by AFt Phase

Birnin-Yauri and Glasser (1998) presented that chloride ions can be also adsorbed by AFt phase even under the condition of low chloride ion concentration. However, no experimental results were obtained to confirm the validity of this assumption. Hirao et al. (2005) studied the chloride binding capacity of AFt phase and found that chloride ions in external solution cannot be bound by AFt phase and the morphology of AFt phase (Figure 3.8) were unchanged after being exposed to chloride solution. Elakneswaran et al. (2009a) reported that the chloride binding capacity of AFt phase lied between Friedel's salt and C-S-H gel, and the interaction between AFt phase and chloride ions belongs to physical adsorption. The differences of chloride concentration in these studies and the reversibility of physical adsorption may be responsible for the different conclusions drawn in these studies. Generally, the AFt phase in cement paste can decompose into Friedel's salt at high concentration of chloride ions (Hirao et al. 2005). However, the content of AFt phase in cement paste is much less than that of C-S-H gel; the chloride binding of AFt phase can be mostly neglected.

3.5 CHLORIDE CONCENTRATE PHENOMENON

In 1993, Japanese researchers Nagataki et al. (1993) immersed cement paste slices with 3 mm thickness in NaCl solution with the same chloride concentration as sea water (0.547 mol/L). The pore solutions of paste

1 μm

(a)   Before immersion

1 μm

(b)   After immersion into 2 mol/L chloride solution

Figure 3.8 Secondary electron images of Aft. (From Hirao et al. 2005.)

samples were extracted by pore solution expression and the chloride concentration after a period time of immersion was determined. The results
showed that the chloride concentration in the expressed pore solution gradually increased with soaking time. After 28 days of immersion, they found
that the chloride concentration of the expressed pore solution was almost
equal to that in soaking solution, while it almost doubled as the immersion
time sequentially increased to 180 days. They defined this phenomenon as
"chloride condensation," and "chloride condensation index" as the ratio
of chloride concentration in the expressed pore solution to that in soaking solution. However, this phenomenon is closer to the physical adsorption due to the EDL formed on the surface of solid phase, totally different

from the definition of condensation in chemistry. In this part, definitions of "chloride concentrate" and "chloride concentration index" are used to replace "chloride condensation" and "chloride condensation index."

Glass et al. (1996) immersed 5-mm-thick cement paste discs in 0.135 mol/L of NaCl solutions and found that the concentration of free chloride ions in pore solution was 30% higher than that of the exposed solution after 28 days of soaking. The release of weakly bonded chloride under high pressure was considered as the reason for higher chloride concentration. Li et al. (2013) investigated the influences of curing time of cement paste, NaCl soaking concentration, soaking time, soaking temperature, and pore solution expression pressure on the free chloride ion concentration in pore solution of cement pastes. They found that the free chloride ion concentration in pore solution of the cement pastes increased with soaking time first, reached a peak, then decreased with soaking time. The "chloride concentration index" decreased apparently with the increase of chloride ion concentration in soaking solutions.

Baroghel-Bouny et al. (2007) tested the total chloride and water-soluble chloride contents of concrete after soaking in 18.2 g/L of NaCl solution for 28, 56, and 90 days. The water-soluble chloride concentration was almost as high as twice the value of soaking solution, and they ascribed this to "chloride concentrate" phenomenon. In some subsequent research, this phenomenon has been repeatedly verified (He 2010; Yuan 2009).

Nagataki et al. (1993) tried to interpret the "chloride concentration index" in pore solution of cement-based materials by using the theory of EDL first. Due to the surface potential of hydration products, the chloride ions in pore solution can be attracted and can further increase the concentration. During the process of pore solution expression experiments, the chloride ions in EDL can be extracted and increase the chloride concentration in the expressed pore solution. After that, Yuan (2009) calculated the thickness of EDL based on Debye formula (Equation 3.5) and concluded that the higher concentration of soaking solution, the thinner the thickness of EDL and the smaller the concentration index.

$$L = k^{-1} = \sqrt{\frac{RT\varepsilon}{2F^2 c_b}} \tag{3.5}$$

where k is Debye constant, c_b is the concentration of bulk solution, ε is the dielectric constant. Based on the results of Yuan, He (2010) assumed the average concentration of chloride ion in diffuse layer and proposed a computational formula to calculate "chloride concentration index" caused by EDL formed within pore structure of cement-based materials.

Based on a proposed EDL model and a given zeta potential, the relationship between chloride concentration of the expressed pore solution and exposure solution was investigated (He et al. 2016). The variation of ratio of average chloride concentration in bulk pore solution and solution in EDL to chloride concentration of bulk pore solution between samples before (N_{tb}) and after expression (N_{ta}) test and changes of pore structure were determined. The calculation results revealed the relationship between the index and zeta potential, as shown in Figure 3.9. This calculation model explained the reason of higher chloride concentration in expressed pore solution, even though some improvements on the formula used to calculate concentration index were still needed.

The existence of "chloride concentrate" phenomenon has been studied for a period of time; the formation of EDL at the solid–liquid interface and especially the zeta potential of EDL are generally considered as the main influential factors of chloride concentrate and chloride concentration index.

In nature, the phenomena of electrization is ubiquitous at the interface of solid and liquid. The contact of an overwhelming majority of dispersion

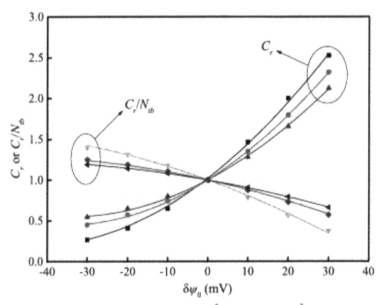

- C_b=0.1 mol/L, y=0.00043x^2+0.038x+1.02, R^2=0.999
- C_b=0.5 mol/L, y=0.00043x^2+0.031x+1.00, R^2=1.000
- C_b=1.0 mol/L, y=0.00037x^2+0.026x+1.00, R^2=0.999
- C_b=0.1 mol/L, y=-0.00013x^2-0.018x+1.00, R^2=0.997
- C_b=0.5 mol/L, y=-0.00010x^2-0.011x+1.00, R^2=1.000
- C_b=1.0 mol/L, y=-0.00008x^2-0.0088x+1.00, R^2=1.000

Figure 3.9 Calculated concentration coefficient (C_r) of chloride ions in the expressed solution and the ratio (C_r/N_{tb}) of concentration coefficient of chloride ions in the expressed pore solution to that capable of transport in the specimen before the expression. (He et al. 2016.)

particles with polar solution or polar media will create an electrical charge. The charged dispersion system will attract ions with opposite charges so that the system as a whole is electrically neutral, which results in the formation of EDL. For solid phases in contact with liquid or immerse in solution, the EDL forms at the interface of solid and liquid due to the ionization, ion exchange or binding and fraction contact.

In 1853, the concept of formation of EDL at the interface of solid and liquid was first proposed by Helmholtz (1853). A parallel plate EDL model was established based on electrostatics theory. However, the thermal motion of ions in liquid was not considered in this model and failed to explain a lot of phenomenon. On the basis of the Helmhollz model, a Gouy–Chapman model was developed and proposed the concept of diffuse layer (Bolt 1955). The ions near the solid surface were under the effects of surface potential and thermal motion, which leads to the decrease of concentration as the distance to surface becomes greater until the same concentration in bulk solution. The Gouy–Chapman model achieved a great breakthrough in understanding the structure of EDL and interpretation of many electro kinetic phenomena. However, this model neglected the radius of ions and used a point charge to express an ion, which resulted in some unreasonable conclusions obtained from this model (Torrie and Valleau 1982). Therefore, Stern (1924) improved the Gouy–Chapman model and proposed a now worldwide-accepted Stern EDL model.

In the EDL model established by Stern, the EDL consists of two layers (Lowke and Gehlen 2017): internal and external layers. Internal layer refers to the compact layer while external layer refers to the diffuse layer. In the compact layer, ions opposite to solid surface potential are strongly bound due to electrical or non-electrical attraction. Out of the compact layer, ions diffusely distribute in diffuse layer, where the concentration of counter-ion (ion with opposite charge of ions in compact layer) is higher than that of ions who take same charge as compact layer (co-ions). With the increase of distance to solid surface, the concentration of counter-ion decreases to that of bulk solution. In diffuse layer, there exists a shear plane, which is located close to the boundary layer between compact and diffuse layer. The solution on the side of pore solution is flowable and able to move with bulk solution. This movement generates a potential difference at the shear plane, which is zeta potential.

3.5.1 Formation of EDL within Cement-Based Materials

For cement-based materials, the EDL forms at the interface between pore wall and pore solution. The main hydration product of cement-based materials is calcium silicate hydrates. In fully hydrated Portland cement paste, C-S-H may reach up to 70% of total volume. Meanwhile, C-S-H has a high specific surface area. The surface of C–S–H is negatively charged due to ionization of silanol sites in alkaline condition. The dissociation equilibrium of silanol sites due to pH increase is given by:

$$\equiv SiOH + OH^- \Leftrightarrow \equiv SiO^- + H_2O \tag{3.6}$$

Within the cement-based materials, the hydration of cement may form portlandite and increase the concentration of calcium ions in pore solution. Generally, the pore solution in cement-based materials is saturated with calcium ions. The adsorption of Ca^{2+} on the silanol sites resulted in a positive charge:

$$\equiv SiOH + Ca^{2+} \Leftrightarrow \equiv SiOCa^+ + H^+ \tag{3.7}$$

When cement-based materials are immersed in a NaCl solution or chloride ions are penetrated into the samples from the environment, the chloride ions in pore solution can be adsorbed onto the silanol sites:

$$\equiv SiOH + Ca^{2+} + Cl^- \Leftrightarrow \equiv SiOCaCl + H^+ \tag{3.8}$$

As the concentration of chloride ions in pore solution increases, the surface potential is decreased due to the adsorption of Cl^-:

$$\equiv SiOH + Cl^- \Leftrightarrow \equiv SiOCl^{2-} + H^+ \tag{3.9}$$

He et al. (2016) suggested the following equation to calculate the coefficient of concentration increase (N_c) in a single-size pore system:

$$N_c = \frac{\Delta\left(L^2 + 2R_iL\right) + R_i^2}{\left(L + R_i\right)^2} \tag{3.10}$$

$$\Delta = \frac{4}{1 - \tanh\left(\dfrac{F}{4RT}\varphi_0\right)\exp(1)} - \frac{4}{1 - \tanh\left(\dfrac{F}{4RT}\varphi_0\right)} + 1 - \exp(-1)$$

where R_i is pore size, ψ_o is Stern plane potential, L is the thickness of EDL, R is gas constant, T is temperature, and F is Faraday constant.

Figure 3.10 shows the relationship between N_c and pore diameter R_i as calculated by Equation 3.10. It can be seen that N_c decreases with the increase of R_i. This trend seems to be more significant when R_i is less than 20 nm and when Stern plane potential is increasing. As R_i further increases, N_c gets close to 1.0.

During the studies on ion transport in cement-based materials, a physical model of the EDL was proposed as shown in Figure 3.11 (Friedmann et al. 2008). Based on the study of Eagland (1975) and Stern EDL model, Lowke

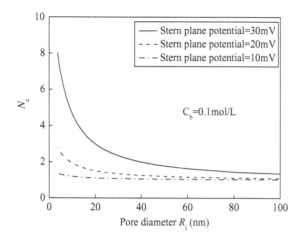

Figure 3.10 Relationship between N_c and R_i. (From He et al. 2016.)

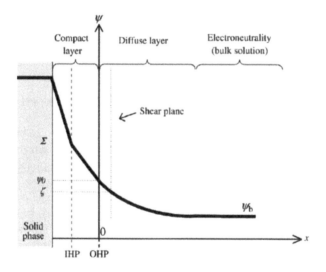

Figure 3.11 Schematic view of the EDL model. (Friedmann et al. 2008.)

and Gehlen (2017) proposed an EDL model close to Figure 3.11; they also determined the potential distribution within the EDL shown in Figure 3.12.

Friedmann et al. (2008) numerically modeled the electrical potential and the chloride ions distribution within diffuse layer of EDL with Poisson–Boltzmann and Nernst–Planck equations. The thickness of the diffuse layer was calculated, and the overlapping of EDL in small pore was also investigated. Based on the proposed physical model and calculation of chloride distribution in EDL, the effects of EDL on ion migration in

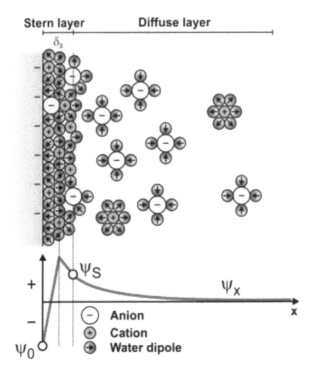

Figure 3.12 Scheme of the EDL and potential distribution. (From Lowke and Gehlen 2017.)

cement-based materials were studied. Nguyen and Amiri (2014) presented that EDL imposed more significant influences on the saturated concrete than unsaturated concrete due to the continuity of liquid phase. The chloride concentration in EDL was affected by the sign of zeta potential in EDL. Generally, the positive zeta potential can increase the chloride concentration in EDL, while the chloride concentration may be lower than that in bulk solution with negative zeta potential. In the study of Nguyen and Amiri (2016), the simulated chloride profiles were compared with the measured results and a good agreement was achieved when the effects of EDL were taken into account with the proposed model. The zeta potential and pore diameter showed significant effects on chloride concentration within EDL, as shown in Figure 3.13. In order to validate the modeling of EDL effects, simulated current densities were compared to the monitoring results of current densities during chloride migration tests on cement mortars and a good agreement was obtained (Friedmann et al. 2012).

3.5.2 Zeta Potential

Generally, zeta potential of suspensions is one of the most relevant parameters controlling the rheological behavior, and extensive studies have been

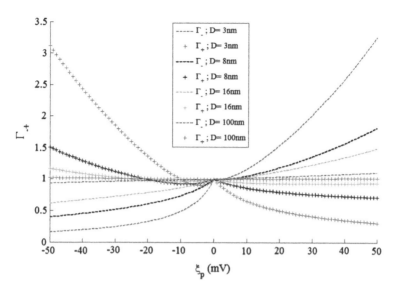

Figure 3.13 Evolution of the "chloride concentration index" according to the characteristics related to the EDL (diameter, Zeta potential). (From Nguyen and Amiri 2016.)

conducted on the zeta potential of cementitious suspensions and freshly mixed cement pastes. In a recent paper (Lowke and Gehlen 2017), the zeta potential of cement and additions in cementitious suspensions with high solid fraction was investigated. The results showed that the value of zeta potential was mainly determined by concentrations of divalent calcium and sulfate ions, and the adsorption of these ions onto the EDL decreased the absolute value of zeta potential. Besides the concentration of ions in solution, pH value and specific surface area of powders were considered to affect the value of zeta potential of suspension (Júnior and Baldo 2014). The studies of Gunasekara et al. (2015) presented that the negative zeta potential of fly ash and other geopolymer indicated more reactive of these materials and more gel formation. Ersoy et al. (2013) studied the influences of mixing water type (pure water, tap water, NaCl and $CaCl_2$ salt water) on zeta potential of Portland cement suspensions with time, and different time-dependent zeta potential values of Portland cement paste were obtained with different mixing water. Due to the relationship between zeta potential and rheology of cement suspension, the effects of superplasticizer on zeta potential and rheological properties of cement paste have been also studied (Liu et al. 2015).

For hydrated cement-based materials, many researches have reported the mechanism of EDL formation at the solid–liquid interface. Elakneswara et al. (2009a) measured the zeta potential of hydrated cement paste and its main phases in water. Friedel's salt and portlandite showed positive surface charge while other phases had negative surfaces in water (Figure 3.14). For

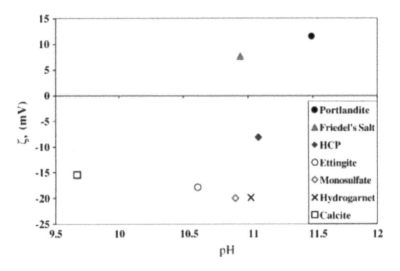

Figure 3.14 Zeta potential of different particles in water. (Elakneswaran et al. 2009a.)

cement-based materials, the silanol sites (SiOH) of C-S-H gel were considered as the main phase controlling the properties of EDL. The physical adsorption of ions in pore solution onto surface of C-S-H can be expressed as Equations 3.7, 3.8, and:

$$\equiv SiOH + Ca^{2+} + SO_4^{2-} \Leftrightarrow \equiv SiOCaSO_4^- + H^+ \tag{3.11}$$

In another paper of Elakneswara et al. (2009b), the zeta potential and adsorbed ions content of slag-blended cement paste were studied. The results showed that the negative zeta potential may gradually increase to positive one with the increase of Ca^{2+} concentration, and both calcium and chloride were potential determining ions. However, the studies of Hocine et al. (2012) showed that a negative zeta potential of cement paste was obtained when sodium hydroxide, potassium hydroxide solutions, or deionized water were used as a support solution, while the addition of sodium chloride tended to make the zeta potential positive (Figure 3.15).

Understanding the properties of EDL and potential distribution in EDL (or diffuse layer) is important to study the mechanism of "chloride concentration" of cement-based materials. Zeta potential, or potential at the solid–liquid interface, is a fundamental parameter characterizing the properties of EDL. Zeta potential is a parameter associated with detailed chemistry and distribution of potential or ions in EDL between substrate and solution. The assumptions in the EDL model and the complex nature of zeta potential make it complicated and difficult to measure and interpret (Kirby and Hasselbrink 2004). Generally, the measurement techniques on zeta potential can be divided into three groups, (1) electrodialytic method,

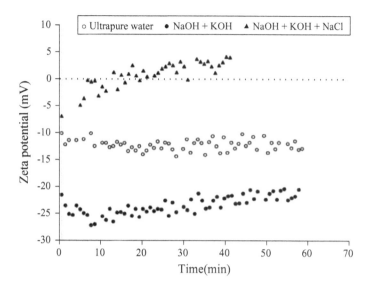

Figure 3.15 Zeta potential evolution of hardened cement paste with testing time. (From Hocine et al. 2012.)

(2) streaming potential method or electrophoresis test, and (3) measuring response of a small spherical particle in an applied E-field or electroacoustic method.

Electrodialytic method is a straightforward method to measure the zeta potential by testing the electroosmotic mobility of the dispersion medium, which can be directly related to zeta potential with the assumption of thin Debye layer. In early papers, the electrodialytic method has been used to measure the zeta potential of single crystals (Smit and Stein 1977). However, the significant effects of accumulating heat with the increasing electric field and surface tension on testing results mostly contribute errors to the measurement results.

Unlike the electrodialytic method, streaming potential method is a method to monitor the motion of dispersion phase relative to a fluid under the influence of an electric field. The most widely accepted theory of electrophoresis was developed in 1903 by Paillot (1904), and the zeta potential-ζ can be determined by the measured electrophoretic mobility-μ

$$\mu = \frac{\zeta \epsilon_m V}{4\pi\eta D} \tag{3.12}$$

where V is the applied voltage, η is the viscosity of the solution, ϵ_m is the dielectric constant of the medium, and D is the electrode separation. The streaming potential method has been widely applied in zeta potential measurement of suspension samples (White et al. 2007), and has also been used in studying the hydration of cement (Nägele 1985, 1986) and interaction between ions and hydration products (Viallis-Terrisse et al. 2001) and also ion interaction with solid phase (He et al. 2016). However, Equation 3.12 is an approximation that only works for thin EDL, and the effects of electrode polarization also contribute to the sources of zeta potential measurement errors by this method, which limits the application of this technique.

In 1933, Debye (1933) found that an alternating electrical field can be generated with an applied sound wave passing through an electrolyte. A macroscopic current of the generated field can be measured by electrokinetic measurement, the current for solid samples was defined as colloid vibration current (CVI) or colloid vibration potential (CVP) (Zana and Yeager 1982). The particle size distribution and zeta potential of colloid samples can be determined with the measured electrophoretic mobility or CVI under applied sound wave. For cement-based materials, electroacoustic method is mainly applied in investigating the hydration (Plank and Hirsch 2007), superplasticizer adsorption (Ferrari et al. 2010; Plank and Sachsenhauser 2006; Zingg et al. 2008), and chloride migration process (Plank and Gretz 2008).

Compared to other methods, such as electrophoretic and electrodialytic techniques, the electroacoustic technique is reported to be superior in zeta potential measurement of aqueous colloids. Acoustic waves are appropriate for complicated system with high concentrations. The electroacoustic method has the advantages of measuring zeta potential of samples with higher concentrations, larger particle size range, consuming fewer samples, and having been used to investigate the electrochemical properties of various materials. In this study, the electroacoustic method is used to measure the zeta potential of freshly mixed and hardened cement paste, which can to some extent avoid variations in hydration and surface properties between diluted and original samples. The mechanism of electroacoustic measurement and calculation of zeta potential can be expressed as follows in the next paragraph.

When placed in a sound field, the stability of EDL may be broken down. The surplus ions are removed and a new polarized state reoccurs, which results in the rearrangement of electric charge at the surface of particles. The polarization effects induce dipole moment and produce a colloid vibration current within the suspension system.

The colloid vibration current (CVI, I_{CV}) is a kind of current at macro level for the suspension system. For thin EDL of monodisperse suspension and low conductivity, the macroscopic current and field intensity can be calculated based on Shilov–Zharkikh cell model:

$$I = I_r / \cos\theta_{r=b} \quad E = \phi / b\cos\theta_{r=b} \tag{3.13}$$

where r is polar coordinate, θ is polar angle, b is the cell diameter in cell model, ϕ is potential (V), and $<E>$ is macroscopic field intensity *(V/m)* and current *(A)*.

Based on Equation 3.13, I_{CV} can be calculated as

$$I_{CV} = \left(-K_m / \cos\theta\right)\left(\partial\phi / \partial r_{r=b}\right) \tag{3.14}$$

where K_m is conductivity of medium (S/m). Introducing the Kuwabara cell model and considering the overlapping of EDL:

$$I_{CV} = \zeta GQ(1+F)\nabla p \tag{3.15}$$

$$G = (9j / 2s)h(s) / \left\{3H + 2s / 1 - \rho_p (1-\varphi) / \rho_s\right\}$$

$$Q = 2\varepsilon_0\varepsilon_m\varphi\left(\rho_p - \rho_s\right) / (3\eta\rho_s), \; F = (1-\varphi) / (2+\varphi), \nabla p$$

where ζ is zeta potential (V), ∇p is pressure gradient (Pa/m), j is imaginary unit, $s = a[\omega / (2\upsilon)]^{1/2}$, a is fineness of dispersion phase (m), ω is ultrasonic frequency (rad/s), υ is kinematic viscosity (m²/s), h, H, S is special function proposed in Kuwabara cell model, ρ_p, ρ_s is density of dispersion phase and colloid respectively (kg/m³), φ is volume concentration, ε_0 and ε_m is vacuum dielectric constant (F/s) and relative dielectric constant of dispersion medium, η is viscosity (Pa·s).

3.5.3 Chloride Distribution within EDL

The ion distribution in the diffuse layer of EDL can be modeled by Gouy–Chapman model; based on the assumption in this model, the probability of ions existing in diffuse layer is proportional to Boltzmann factor $e^{-ze\psi/kT}$ The ion concentration in diffuse layer can be expressed as follow:

$$n_+ = n_0\exp\left(-\frac{ze\psi}{kT}\right)$$

$$n_- = n_0\exp\left(\frac{ze\psi}{kT}\right) \tag{3.16}$$

where

n_+, n_- The number of cation or anion per unit volume solution

n_0 The number of ions per unit volume solution at the distance where electric neutrality($\psi = 0$)is reached

z Charge number of cation or anion

ψ The potential at distance x from shear plane

According to classical theory of electrostatics, the relationship of potential in diffuse layer and the distance from solid surface (or shear plane) can be expressed by Poisson equation.

$$\nabla^2 \psi = -\frac{\rho}{\varepsilon_0 \varepsilon_r} = -\frac{ze(n_+ - n_-)}{\varepsilon_0 \varepsilon_r} = \frac{2n_0 ze}{\varepsilon_0 \varepsilon_r} \sinh\left(\frac{ze\psi}{kT}\right) \tag{3.17}$$

where

$\varepsilon_0, \varepsilon_r$ Vacuum permittivity of EDL and relative permittivity of solution

∇_2 Laplace operator, which express the divergence of a function

ρ Charge balance, which is the number of negative charge minus that of positive charge per unit volume

Solving the equation above with boundary conditions introduced, the following result is obtained:

$$\kappa x = \ln \frac{\left(e^{\frac{y}{2}} + 1\right)\left(e^{\frac{y0}{2} - 1} - 1\right)}{\left(e^{\frac{y}{2}} - 1\right)\left(e^{\frac{y0}{2} - 1} + 1\right)} \tag{3.18}$$

where $\kappa^2 = \dfrac{2n_0 z^2 e^2}{\varepsilon_0 \varepsilon_r kT}, \dfrac{1}{\kappa}$ is defined as effective thickness of diffused EDL:

$y = ze\psi / kT$, and $y0$ is equal to the value of y at $x = 0$.

Considering the value of y_0, in our experiment, the maximum value of zeta potential equals 7.87 mV. Generally, the zeta potential of cement-based materials is at the range between −20 to 20 mV. Introduced into the equation above, it can be known that the values of y are all lower than 1.0. By series expansion of $e^{y/2}$ and leaving the former two terms, it holds that:

$$e^{y/2} = 1 + \frac{y}{2} + \frac{(y/2)^2}{2!} + \frac{(y/2)^3}{3!} + \ldots \approx 1 + y/2 \tag{3.19}$$

Substitution into Equation 3.18 leads to:

$$\kappa x = \ln \frac{(2 + y/2) \times y_0/2}{y/2 \times (2 + y_0/2)} \approx \ln \frac{y_0}{y} = \ln \frac{\psi_0}{\psi} \text{ or } \psi = \psi_0 e^{-\kappa x} \tag{3.20}$$

Substituting Equation 3.20 and effective thickness of EDL into Equation 3.16, the concentration of chloride ions c_{Cl} at distance x from shear place is obtained by:

$$c_{Cl} = n_0 \exp\left(\frac{ze\psi_0 e^{-\kappa x}}{\kappa T}\right) = n_0 \exp\left(\frac{ze\psi_0 e^{-x\left(\frac{2n_0 z^2 e^2}{\varepsilon_0 \varepsilon_r \kappa T}\right)^{1/2}}}{\kappa T}\right) \tag{3.21}$$

Integrating the chloride concentration in the diffuse layer, we can obtain the average concentration of chloride ions as follows:

$$n_{ave} = \kappa \int_0^L n_0 \exp\left(\frac{ze\psi_0 e^{-x\left(\frac{2n_0 z^2 e^2}{\varepsilon_0 \varepsilon_r \kappa T}\right)^{1/2}}}{\kappa T}\right) dx \tag{3.22}$$

With the assumption of cylindrical pore model (Figure 3.16), the chloride concentration in the expressed pore solution can be obtained:

$$N_c = \int \frac{n_{ave} \times \pi\left[(R+L)^2 - R^2\right] + n_0 \times \pi R^2}{n_0 \times \pi (R+L)^2} d(R+L) \tag{3.23}$$

Figure 3.17 shows the results of N_c of cement paste with different slag replacement level (0, 20%, 40%, and 60%) obtained from pore solution

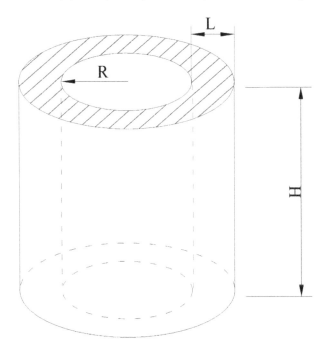

Figure 3.16 Pore structure model within cement-based materials.

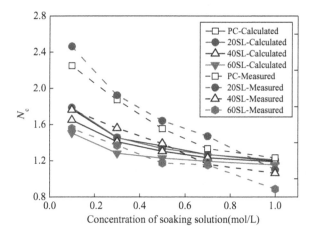

Figure 3.17 Comparison of calculated and measured chloride concentration index N_c. (From Hu 2017.)

expression and calculated according to the equations shown above and the measured zeta potential by electroacoustic method (Hu 2017). It can be seen that the development of calculated concentration index with chloride concentration in soaking solution and slag replacement was in agreement with the measured results. However, some differences can also be investigated. The fact that we assumed the pore shape as a simplified cylindrical model with a simple pore diameter distribution and the higher value of surface potential than zeta potential may lead to the underestimate of the calculated results.

3.5.4 Variations of EDL during Pore Solution Expression

Water extraction and pore solution expression methods are widely accepted and applied for the acquisition of pore solution and the determination of free chloride concentration. The results obtained by using water extraction methods are considered as water-soluble chloride content, which is mostly higher than free chloride concentration when high water-solid ratio and long extraction time is applied. Pore solution expression is regarded as the most reliable method to study the free chloride concentration and has been widely applied. The validity of pore solution expression method has been evaluated in some studies (Duchesne and Bérubé 1994), and it was shown that the expressed pressure did not impact any effects on the ion concentration in the expressed pore solution. However, when high pressure is applied to cement-based materials samples, chloride ions in diffuse layer will also be extracted and will increase the chloride concentration of the expressed pore solution. In the research where chloride concentrate phenomenon was

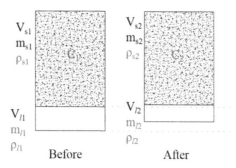

V_{s1}
m_{s1}
ρ_{s1}

V_{l1}
m_{l1}
ρ_{l1}

V_{s2}
m_{s2}
ρ_{s2}

V_{l2}
m_{l2}
ρ_{l2}

Before   After

Figure 3.18 Schematic diagram of samples before and after pore solution expression.

reported, the pore solution expression method was generally applied to obtain the pore solution of cement-based materials.

However, pore solution expression method also has many deficiencies, which to some extents limits the application of this method in studying ion migration and interaction between ions and hydration products of cement-based materials. First, special expression apparatus and high pressure is needed, and only a bit of pore solution can be obtained for every test. The dilution in the obtained solution may affect the accuracy of the ion concentration test. Meanwhile, it is not sure whether all of the pore solution within samples can be extracted or which part of the pore solution can be extracted. The determination of effects of pore solution expression on microstructure and chloride binding of cement-based materials are also challenges for further studies.

Figure 3.18 is the schematic diagram of cement paste before and after pore solution expression experiment. During the pore solution expression, part of pore solution within cement paste can be extracted, while the content of solid phase can be considered unchanged. For the expressed pore solution, the amount of chloride ions n_{Cl} can be calculated as:

$$n_{Cl} = (C_1 - C_2) * m_s / 35.45 \tag{3.24}$$

where C_1 and C_2 is total chloride content of pastes before and after pore solution expression, m_s is the mass of solid phase of cement paste. Then, the volume of pore solution (V_p) expressed by pore solution expression is:

$$Vp = \frac{m_{l1}}{\rho_{l1}} - \frac{m_{l2}}{\rho_{l2}} = \frac{m_s \omega_1}{(1 - \omega_1)\rho_{l1}} - \frac{m_s \omega_2}{(1 - \omega_2)\rho_{l2}} \tag{3.25}$$

where m_{l1} and m_{l2}, ρ_{l1} and ρ_{l2} is mass and density of pore solution before and after pore solution expression. Generally, the density of immersed NaCl solution can be used as the density of pore solution ρ_l. Then, the chloride concentration in the expressed pore solution can be calculated as:

$$N_c = n_{Cl} / Vp = (C_1 - C_2) / \left\{ 35.45 \times \left[\frac{m_s \omega_1}{(1 - \omega_1)\rho_{l1}} - \frac{m_s \omega_2}{(1 - \omega_2)\rho_{l2}} \right] \right\} \quad (3.26)$$

In our studies on microstructure variations of cement paste during pore solution expression process, the pore solution distribution and chloride contents of samples before and after pore solution expression were tested by ^{1}H NMR and total chloride content tests. Nuclear magnetic resonance (NMR) is a physical phenomenon with the absorption and re-emitting of electromagnetic radiation by a nuclei in a magnetic field. Depending on the magnetic field strength and magnetic properties of the nuclei, the absorption and re-emitting energy at a specific resonance frequency can provide some information on molecular structure and multi-phase interaction of crystal and non-crystalline materials.

Water is an essential reactant of cement hydration and important media of aggressive substances, in which ^{1}H NMR measurement can be used to detect the content and distribution results of cement-based materials (Friedemann et al. 2006; Greener et al. 2000). By measuring the ^{1}H NMR spectrum of cementitious materials during hydration, extensive studies have been conducted on the cement hydration process (Cano-Barrita et al. 2009; Muller et al. 2013; Puertas et al. 2004), porosity, and pore size distribution determination (Pipilikaki and Beazi-Katsioti 2009; Kupwade-Patil et al. 2018) by ^{1}H NMR. Different types of water including free state water and adsorbed water simultaneously coexist within cement-based materials, and characterization of water in different pore size ranges, such as intra- and inter- C-S-H gel pore, is of great significance for durability studies of cement-based materials, which has been also studied in previous studies applying ^{1}H NMR measurement (McDonald et al. 2010).

From the ^{1}H NMR test, signal amplitude of water but not the water mass within samples can be obtained. Thus, the relationship between water mass and signal amplitude were established by running a series of ^{1}H NMR on a specimen with different water content. For all of the cement pastes prepared in our studies, the correlations between NMR signal amplitude and the mass of measured water within samples are obtained and shown in Figure 3.19. It can be seen that the correlation between the signal amplitude and the mass of water is governed by w/b ratio and pore solution expression process. According to the slope of fitting line in Figure 3.19 and the tested signal amplitude before and after pore solution expression without drying, the mass ratio of water within paste can be obtained.

Figure 3.20 is T2 relaxation distribution plot of cement paste with different w/b ratios, and results of pastes before and after pore solution expression

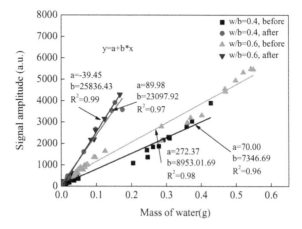

Figure 3.19 Relationship between tested water mass and signal amplitude. (From Hu 2017.)

experiment are all presented. The T2 relaxation time was transformed to pore size, shown in x axis (nm) in every plot in Figure 3.20. Before pore solution expression, it can be seen clearly from the figures that the cement paste sample with w/b ratio 0.6 has a larger porosity than that of 0.4. With the decrease of w/b ratio, the amplitude of signal was reduced and the T2 relaxation time or pore size of the peak shifted left. The effects of w/b ratio on porosity and pore structure of cement paste obtained from NMR are in agreement with the results from MIP or other test methods shown in Ding and Li (2005) and Moon et al. (2006). After the pore solution expression, it can be seen that samples with different w/b ratios showed less differences for water distribution or pore size distribution. The pores with diameter larger than 40nm were all mostly removed from samples after the pore solution expression, while the porosity of smaller pores was decreased.

Together with the total chloride results, the chloride concentration and chloride concentration index N_c cab be calculated. It can be seen from Figure 3.21 that the chloride concentration in the expressed pore solution is higher than that in soaking solution since N_c of all pastes are larger than 1.0. With the increase of chloride concentration in soaking solution and w/b ratio, the value of N_c is obviously decreased, which is in agreement with the results measured from pore solution expression. The comparison between measured and calculated N_c is also shown in Figure 3.21. It can be seen from the figure that similar trends of N_c with chloride concentration and w/b ratio are obtained between the measured and calculated results. However, the results from the calculation are lower than that of measured ones. Due to limited pore solution available for analysis, the calculation results are very sensitive to the data from NMR test and total chloride content. It can be imagined that part of chloride ions, which should be extracted out, may

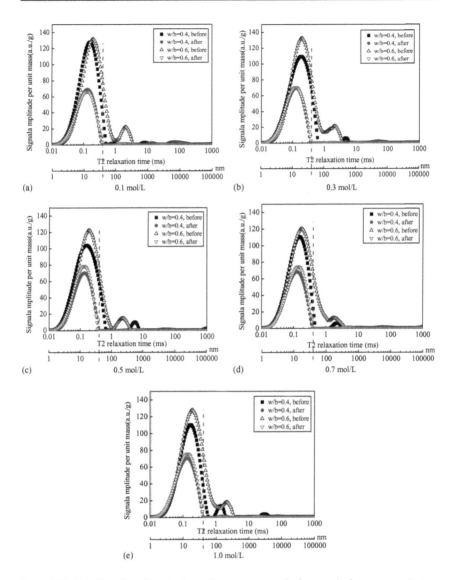

Figure 3.20 T_2 relaxation distribution of cement paste before and after pore solution expression. (From Hu 2017.)

remain on the surface of pastes after pore solution expression. However, during our experiment, the powder samples after pore solution were not washed or flushed to prevent the loss of water-soluble chloride. Therefore, this part of chloride was considered as the chloride ions remained within samples after pore solution expression. This may result in lower calculated amount of chloride in the expressed pore solution and thus lower N_c.

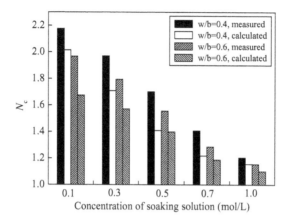

Figure 3.21 Comparison of Nc between measured and calculated results. (From Hu 2017.)

3.6 CHLORIDE BINDING WITH OTHER COMPOUNDS

Smolczyk (1968) found that the concrete immersed in 3 mol/L $CaCl_2$ solution was cracked due to an expanding reaction. The products of the expanding reaction was verified as $CaCl_2 \cdot Ca(OH)_2 \cdot H_2O$. All the calcium hydroxide disappeared in cement paste stored in this solution. Trætteberg (1977) investigated the hydration products of cement pastes immersed in NaCl and $CaCl_2$ solution by DTA/TG and XRD. The content of bound chloride ions WAS greater for samples immersed in NaCl solution than that in $CaCl_2$ solution. The $Ca(OH)_2$ content in $CaCl_2$-immersed samples decreased and almost totally disappeared after one year of immersion. Lambert et al. (1985) compared the chloride binding capacity of alite (C_3S) phase immersed in NaCl and $CaCl_2$ solution; the results showed that no chloride ions can be bounded by C_3S phase in $CaCl_2$ solution. Lambert et al. (1985) verified that when $CaCl_2$ was applied as chloride salt in $CaCl_2$-$Ca(OH)_2$-H_2O system, $CaCl_2 \cdot 3Ca(OH)_2 \cdot 12H_2O$ was formed but not $CaCl_2 \cdot Ca(OH)_2 \cdot H_2O$, which can be represented by Equation 3.27:

$$CaCl_2 + 3Ca(OH)_2 + 12H_2O = CaCl_2 \cdot 3Ca(OH)_2 \cdot 12H_2O \qquad (3.27)$$

Shi (2001) reported that $CaCl_2 \cdot 3Ca(OH)_2 \cdot 12H_2O$ was very unstable and could decompose under the condition of 20% relative humidity and 20°C temperature. Therefore, the condition of experiment is strictly required to be controlled for identifying $CaCl_2 \cdot 3Ca(OH)_2 \cdot 12H_2O$ in cement-based materials.

Hirao et al. (2005) immersed the analytically pure $Ca(OH)_2$ in 2 mol/L NaCl solution, as shown in Figure 3.22; the morphology of $Ca(OH)_2$ was unchanged before and after immersion in NaCl solution. The XRD results also showed that no new phase was formed in the system. However,

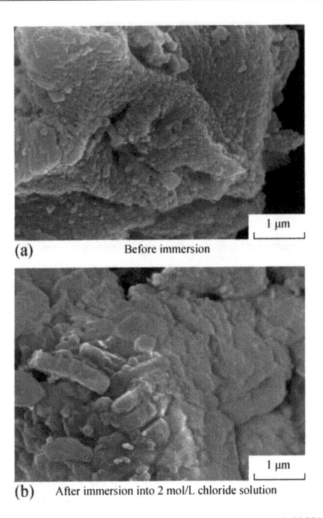

(a)   Before immersion

(b)   After immersion into 2 mol/L chloride solution

Figure 3.22 Secondary electron images of Ca(OH)$_2$. (From Hirao et al. 2005.)

Elakneswaran et al. (2009a) found that the ionization of Ca(OH)$_2$ resulted in positive charge of the surface and the ability to physically adsorb chloride ions onto the surface of Ca(OH)$_2$ and form CaOHCl. The crystal structure of CaOHCl was also detected by XRD.

Elakneswaran et al. (2009a) detected the bound chloride of synthetic portlandite and Friedel's salt and obtained the chloride binding isotherms shown in Figure 3.23. The mechanism of chloride binding by portlandite and Friedel's salt can be studied and expressed with the equations below:

$$\left[Ca_2Al(OH)_6 \right]^+ + Cl^- + 2H_2O \Leftrightarrow \left[Ca_2Al(OH)_6 \right] Cl \cdot 2H_2O \qquad (3.28)$$

$$\equiv \left[CaOH \right]^+ + Cl^- \equiv CaOHCl \qquad (3.29)$$

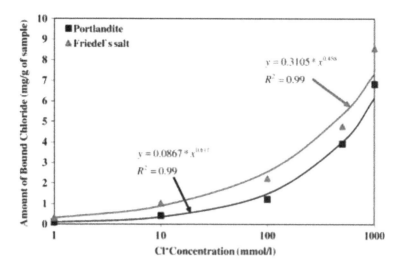

Figure 3.23 Amount of adsorbed chloride vs. initial chloride concentration for portlandite and Friedel's salt. (From Elakneswaran et al. 2009a.)

The abilities of portlandite and Friedel's salt to bind chloride were ascribed
to the physical adsorption of chloride ions onto the surface of solid phase.

According to Elakneswaran et al. (2009a), ettringite can also bind chloride ions by physical adsorption, while the remarkably small amount of
ettringite in cement-based materials to that of C-S-H made this effect negligible. Even though some studies (Hirao et al. 2005) obtained results that
no chloride binding capacity can be obtained by ettringite, we can still
conclude that ettringite possesses the ability to bind chloride at some specific chloride concentration range, and monosulfate plays an important role
(Balonis et al. 2010).

3.7 SUMMARY

Chloride binding plays an important role in the chloride penetration process and service life prediction of cement-based materials. Chloride can be
presented within cement-based materials in the form of free state, physically adsorbed, and chemically bound chloride. Generally, the corrosion
of reinforcing steel is mostly affected by free state chloride in pore solution. "Chloride concentrate" is a phenomenon that chloride concentration
in expressed pore solution from pore solution expression test is higher than
that in soaking solution. Physical adsorption from EDL is considered as the
reason for this phenomenon.

This chapter reviews the mechanism of chemical binding and physical
adsorption; the research progress of formation and stability of Friedel's

salt and the "chloride concentrate" phenomenon are discussed in detail to explain the mechanism of chemical binding and physical adsorption.

The following conclusions can be made:

(1) Chloride ions can be internally mixed or externally intruded into concrete, in which formation of Friedels's salt is the main mechanism of chemical binding. However, the formation of Friedel's salt is different for internal or external chloride. When chloride ions are internally mixed, the formation of Friedel's salt mainly comes from the interaction between chloride ions and C_3A with the dissolution and precipitation mechanism, while for external chloride ions, ion exchange is the main mechanism of the reaction between chloride ions and AFm phase.

(2) The stability of Friedel's salt has a dependency on chloride concentration and pH value of pore solution. When chloride ions are penetrated or dissolved in pore solution, Kuzel's salt is first formed at low chloride concentration, then transforms into Friedel's salt. Friedel's salt can also decompose to Kuzel's salt when chloride concentration of pore solution decreases. Further study is still needed to make it clear whether Kuzel's salt will be formed under the low concentration of internal chloride ions.

(3) The chloride binding capacity of hydration products is different; the AFm phase comes first, then C-S-H gel. The former interacts with chloride ions by chemical binding, while the reaction between the latter and chloride ions is physical adsorption. AFt phase, $Ca(OH)_2$, and Friedel's salt can also interact with chloride ions, which can be neglected due to small content in cement-based materials and low chloride binding capacity.

(4) "Chloride concentrate" phenomenon is confirmed for cement-based materials, and the higher chloride concentration in the expressed pore solution results from the attraction of EDL on chloride ions.

(5) EDL models, especially the developed Stern model, can be used to describe the potential and ion distribution within EDL; zeta potential is an important parameter to characterize the properties of EDL.

REFERENCES

Balonis M, Lothenbach B, Le Saout G, Glasser FP. Impact of chloride on the mineralogy of hydrated Portland cement systems. *Cement and Concrete Research*. 2010, 40:1009–1022.

Baroghel-Bouny V, Belin P, Maultzsch M, Henry D. $AgNO_3$ spray tests: Advantages, weaknesses, and various applications to quantify chloride ingress into concrete. Part 1: Non-steady-state diffusion tests and exposure to natural conditions. *Materials and Structures*. 2007, 40:759.

Baroghel-Bouny V, Wang X, Thiery M, et al. Prediction of chloride binding isotherms of cementitious materials by analytical model or numerical inverse analysis. *Cement and Concrete Research*. 2012, 42(9):1207–1224.

Ben-Yair M. The effect of chlorides on concrete in hot and arid regions. *Cement and Concrete Research.* 1974, 4:405–416.

Birnin-Yauri U, Glasser F. Friedel's salt, Ca_2Al (OH) 6 (Cl, OH) $\cdot$ $2H_2O$: Its solid solutions and their role in chloride binding. *Cement and Concrete Research.* 1998, 28:1713–1723.

Bolt G. Analysis of the validity of the Gouy-Chapman theory of the electric double layer. *Journal of Colloid Science.* 1955, 10:206–218.

Cano-Barrita PDJ, Marble A, Balcom B, et al. Embedded NMR sensors to monitor evaporable water loss caused by hydration and drying in Portland cement mortar. *Cement and Concrete Research.* 2009, 39:324–328.

Chen Y, Shui Z, Chen W, et al. Chloride binding of synthetic $Ca–Al–NO_3$ LDHs in hardened cement paste. *Construction and Building Materials.* 2015, 93:1051–1058.

Csizmadia J, Balázs G, Tamás FD. Chloride ion binding capacity of aluminoferrites. *Cement and Concrete Research.* 2001, 31:577–588.

Debye P. A method for the determination of the mass of electrolytic ions. *The Journal of Chemical Physics.* 1933, 1:13–16.

Diamond S. Chloride concentrations in concrete pore solutions resulting from calcium and sodium chloride admixtures. *Cement, Concrete and Aggregates.* 1986, 8:97–102.

Diamond S, Dolch W, White JL. Studies on tobermorite-like calcium silicate hydrates. Highway Research Record, 1964.

Ding Z, Li Z. Effect of aggregates and water contents on the properties of magnesium phospho-silicate cement. *Cement and Concrete Composites.* 2005, 27:11–18.

Duchesne J, Bérubé M. Evaluation of the validity of the pore solution expression method from hardened cement pastes and mortars. *Cement and Concrete Research.* 1994, 24:456–462.

Eagland D. The influence of hydration on the stability of hydrophobic colloidal systems. *Water in Disperse Systems.* Springer; 1975, 1–74.

Ekolu S, Thomas M, Hooton R. Pessimum effect of externally applied chlorides on expansion due to delayed ettringite formation: Proposed mechanism. *Cement and Concrete Research.* 2006, 36:688–696.

Elakneswaran Y, Nawa T, Kurumisawa K. Electrokinetic potential of hydrated cement in relation to adsorption of chlorides. *Cement and Concrete Research.* 2009a, 39:340–344.

Elakneswaran Y, Nawa T, Kurumisawa K. Zeta potential study of paste blends with slag. *Cement and Concrete Composites.* 2009b, 31:72–76.

Ersoy B, Dikmen S, Uygunoğlu T, İçduygu MG, Kavas T, Olgun A. Effect of mixing water types on the time-dependent zeta potential of Portland cement paste. *Science and Engineering of Composite Materials.* 2013, 20:285–292.

Feng W, Xu J, Chen P, et al. Influence of polycarboxylate superplasticizer on chloride binding in cement paste. *Construction and Building Materials.* 2018, 158:847–854.

Ferrari L, Kaufmann J, Winnefeld F, Plank J. Interaction of cement model systems with superplasticizers investigated by atomic force microscopy, zeta potential, and adsorption measurements. *Journal of Colloid and Interface Science.* 2010, 347:15–24.

Florea M, Brouwers H. Chloride binding related to hydration products: Part I: Ordinary Portland cement. *Cement and Concrete Research*. 2012, 42:282–290.

Friedmann H, Amiri O, Aït-Mokhtar A. Physical modeling of the electrical double layer effects on multispecies ions transport in cement-based materials. *Cement and Concrete Research*. 2008, 38:1394–1400.

Friedmann H, Amiri O, Aït-Mokhtar A. Modelling of EDL effect on chloride migration in cement-based materials. *Magazine of Concrete Research*. 2012, 64:909–917.

Friedemann K, Stallmach F, Kärger J. NMR diffusion and relaxation studies during cement hydration—A non-destructive approach for clarification of the mechanism of internal post curing of cementitious materials. *Cement and Concrete Research*. 2006, 36:817–826.

Glass G, Wang Y, Buenfeld N. An investigation of experimental methods used to determine free and total chloride contents. *Cement and Concrete Research*. 1996, 26:1443–1449.

Glasser F, Kindness A, Stronach S. Stability and solubility relationships in AFm phases: Part I. Chloride, sulfate and hydroxide. *Cement and Concrete Research*. 1999, 29:861–866.

Glasser FP. Role of chemical binding in diffusion and mass transport. *Ion and Mass Transport in Cement Based Materials*. American Ceramic Society Special Publication, 2001.

Gong N, Wang X. Examining the binding mechanism of chloride ions in sea sand mortar. 2nd International Conference on Architectural, Civil and Hydraulics Engineering (ICACHE'16). Atlantis Press, 2016.

Greener J, Peemoeller H, Choi C, et al. Monitoring of hydration of white cement paste with proton NMR spin–spin relaxation. *Journal of the American Ceramic Society*. 2000, 83:623–627.

Gunasekara C, Law DW, Setunge S, Sanjayan JG. Zeta potential, gel formation and compressive strength of low calcium fly ash geopolymers. *Construction and Building Materials*. 2015, 95:592–599.

Hassan Z. *Binding of External Chloride by Cement Pastes*. University of Toronto, 2001.

He F. *Measurement of Chloride Migration in Cement-based Materials Using AgNO3 Colorimetric Method*. Central South University, Changsha; 2010.

He F, Shi C, Hu X, et al. Calculation of chloride ion concentration in expressed pore solution of cement-based materials exposed to a chloride salt solution. *Cement and Concrete Research*. 2016, 89:168–176.

Helmholtz P. XLVIII. On the methods of measuring very small portions of time, and their application to physiological purposes. *Philosophical Magazine Series 4*. 1853, 6:313–325.

Hewlett P. *Lea's Chemistry of Cement and Concrete*. Elsevier; 2003.

Hirao H, Yamada K, Takahashi H, Zibara H. Chloride binding of cement estimated by binding isotherms of hydrates. *Journal of Advanced Concrete Technology*. 2005, 3:77–84.

Hocine T, Amiri O, Ait-Mokhtar A, Pautet A. Influence of cement, aggregates and chlorides on zeta potential of cement-based materials. *Advances in Cement Research*. 2012, 24:337–348.

Hu X. *Mechanism of Chloride Concentrate and its Effects on Microstructure and Electrochemical Properties of Cement-Based Materials*. Ghent University; 2017.

Hu X, Shi C, Yuan Q, et al. Influences of chloride immersion on zeta potential and chloride concentration index of cement-based materials. *Cement and Concrete Research*. 2018, 106:49–56.

Júnior JAA, Baldo JB. The behavior of zeta potential of silica suspensions. *New Journal of Glass and Ceramics*. 2014, 4:29.

Ke X, Bernal S A, Provis J L. Uptake of chloride and carbonate by Mg-Al and Ca-Al layered double hydroxides in simulated pore solutions of alkali-activated slag cement. *Cement and Concrete Research*. 2017, 100:1–13.

Kim MJ, Kim KB, Ann KY. The influence of C3A content in cement on the chloride transport. *Advances in Materials Science and Engineering*. 2016, 2016.

Kirby BJ, Hasselbrink EF. Zeta potential of microfluidic substrates: 1. Theory, experimental techniques, and effects on separations. *Electrophoresis*. 2004, 25:187–202.

Kupwade-Patil K, Palkovic S D, Bumajdad A, et al. Use of silica fume and natural volcanic ash as a replacement to Portland cement: Micro and pore structural investigation using NMR, XRD, FTIR and X-ray microtomography. *Construction and Building Materials*. 2018, 158:574–590.

Laidler K, Meiser J. *Physical Chemistry*. The Benjamin Cummings Publishing Company, San Francisco, CA; 1982.

Lambert P, Page C, Short N. Pore solution chemistry of the hydrated system tricalcium silicate/sodium chloride/water. *Cement and Concrete Research*. 1985, 15:675–680.

Larsen C. *Chloride Binding in Concrete, Effect of Surrounding Environment and Concrete Composition*. Trondheim, Norway; 1998.

Li Q, Shi C, He F, et al. Factors influencing free chloride ion condensation in cement-based materials. *Journal of the Chinese Ceramic Society*. 2013, 41:320–327.

Liu M, Lei J, Bi Y, Du X, Zhao Q, Zhang X. Preparation of polycarboxylate-based superplasticizer and its effects on zeta potential and rheological property of cement paste. Journal of Wuhan University of Technology-Mater. Sci. Ed. 2015, 30:1008–1012.

Lowke D, Gehlen C. The zeta potential of cement and additions in cementitious suspensions with high solid fraction. *Cement and Concrete Research*. 2017, 95:195–204.

Ma B, Zhang T, Tan H, et al. Effect of TIPA on chloride immobilization in cement-fly ash paste. *Advances in Materials Science and Engineering*. 2018, 2018.

Matschei T, Lothenbach B, Glasser F. The AFm phase in Portland cement. *Cement and Concrete Research*. 2007, 37:118–130.

McDonald PJ, Rodin V, Valori A. Characterisation of intra-and inter-C–S–H gel pore water in white cement based on an analysis of NMR signal amplitudes as a function of water content. *Cement and Concrete Research*. 2010, 40:1656–1663.

Mehta P. Effect of cement composition on corrosion of reinforcing steel in concrete. *Chloride Corrosion of Steel in Concrete*. ASTM International; 1977.

Midgley H, Illston J. The penetration of chlorides into hardened cement pastes. *Cement and Concrete Research*. 1984, 14:546–558.

Monteiro P, Wang K, Sposito G, Dos Santos M, de Andrade WP. Influence of mineral admixtures on the alkali-aggregate reaction. *Cement and Concrete Research*. 1997, 27:1899–1909.

Moon HY, Kim HS, Choi DS. Relationship between average pore diameter and chloride diffusivity in various concretes. *Construction and Building Materials*. 2006, 20:725–732.

Muller A, Scrivener K, Gajewicz A, McDonald P. Use of bench-top NMR to measure the density, composition and desorption isotherm of C–S–H in cement paste. *Microporous and Mesoporous Materials*. 2013, 178:99–103.

Nägele E. The zeta-potential of cement. *Cement and Concrete Research*. 1985, 15:453–462.

Nägele E. The Zeta-potential of cement: Part II: Effect of pH-value. *Cement and Concrete Research*. 1986, 16:853–863.

Nagataki S, Otsuki N, Wee T-H, Nakashita K. Condensation of chloride ion in hardened cement matrix materials and on embedded steel bars. *Materials Journal*. 1993, 90:323–332.

Nguyen P, Amiri O. Study of electrical double layer effect on chloride transport in unsaturated concrete. *Construction and Building Materials*. 2014, 50:492–498.

Nguyen P, Amiri O. Study of the chloride transport in unsaturated concrete: Highlighting of electrical double layer, temperature and hysteresis effects. *Construction and Building Materials*. 2016, 122:284–293.

Paillot R. M. SMOLUCHOWSKI.—Contribution à la théorie de l'endosmose électrique et de quelques phénomènes corrélatifs (Bulletin de l'Académie des Sciences de Cracovie; mars 1903). *Journal of Physics: Theories and Applications*. 1904, 3:912.

Paul G, Boccaleri E, Buzzi L, et al. Friedel's salt formation in sulfoaluminate cements: A combined XRD and 27Al MAS NMR study. *Cement and Concrete Research*. 2015, 67:93–102.

Pipilikaki P, Beazi-Katsioti M. The assessment of porosity and pore size distribution of limestone Portland cement pastes. *Construction and Building Materials*. 2009, 23:1966–1970.

Plank J, Gretz M. Study on the interaction between anionic and cationic latex particles and Portland cement. *Colloids and Surfaces A: Physicochemical and Engineering Aspects*. 2008, 330:227–233.

Plank J, Hirsch C. Impact of zeta potential of early cement hydration phases on superplasticizer adsorption. *Cement and Concrete Research*. 2007, 37:537–542.

Plank J, Sachsenhauser B. Impact of molecular structure on zeta potential and adsorbed conformation of α-allyl-ω-methoxypolyethylene glycol-maleic anhydride superplasticizers. *Journal of Advanced Concrete Technology*. 2006, 4:233–239.

Puertas F, Fernández-Jiménez A, Blanco-Varela M. Pore solution in alkali-activated slag cement pastes. Relation to the composition and structure of calcium silicate hydrate. *Cement and Concrete Research*. 2004, 34:139–148.

Ramachandran VS. Possible states of chloride in the hydration of tricalcium silicate in the presence of calcium chloride. *Matériaux et Construction*. 1971, 4:3–12.

Saikia N, Kato S, Kojima T. Thermogravimetric investigation on the chloride binding behaviour of MK–lime paste. *Thermochimica Acta*. 2006, 444:16–25.

Shi C. Formation and stability of $3CaO \cdot CaCl_2 \cdot 12H_2O$. *Cement and Concrete Research*. 2001, 31:1373–1375.

Shi Z, Geiker M R, Lothenbach B, et al. Friedel's salt profiles from thermogravimetric analysis and thermodynamic modelling of Portland cement-based mortars exposed to sodium chloride solution. *Cement and Concrete Composites*. 2017, 78:73–83.

Smit W, Stein HN. Electroosmotic zeta potential measurements on single crystals. *Journal of Colloid and Interface Science*. 1977, 60:299–307.

Smolczyk HG. Chemical reactions of strong chloride-solution with concrete. Proceedings of the 5th International Congress on the Chemistry of Cement, 1968, 274–280.

Spice J E. *Chemical Binding and Structure*. Elsevier; 2016.

Stern O. Theory of the electrical double layer (in German). *Electrochemistry*. 1924, 30:508–516.

Suryavanshi A, Scantlebury J, Lyon S. Mechanism of Friedel's salt formation in cements rich in tri-calcium aluminate. *Cement and Concrete Research*. 1996, 26:717–727.

Suryavanshi A, Swamy RN. Stability of Friedel's salt in carbonated concrete structural elements. *Cement and Concrete Research*. 1996, 26:729–741.

Tang LP, Nilsson L-O. Chloride binding capacity and binding isotherms of OPC pastes and mortars. *Cement and Concrete Research*. 1993, 23:247–253.

Thomas M, Hooton R, Scott A, Zibara H. The effect of supplementary cementitious materials on chloride binding in hardened cement paste. *Cement and Concrete Research*. 2012, 42:1–7.

Torrie G, Valleau J. Electrical double layers. 4. Limitations of the Gouy-Chapman theory. *The Journal of Physical Chemistry*. 1982, 86:3251–3257.

Traetteberg A. The mechanism of chloride penetration in concrete. SINTEF report STF65 A, 1977, 77070.

Viallis-Terrisse H, Nonat A, Petit J-C. Zeta-potential study of calcium silicate hydrates interacting with alkaline cations. *Journal of Colloid and Interface Science*. 2001, 244:58–65.

Vieille L, Rousselot I, Leroux F, Besse J-P, Taviot-Guého C. Hydrocalumite and its polymer derivatives. 1. Reversible thermal behavior of Friedel's salt: A direct observation by means of high-temperature in situ powder X-ray diffraction. *Chemistry of Materials*. 2003, 15:4361–4368.

Wang S, Huang Y, Wang Z. Concrete resistance to chloride ingress: Effect of cement composition. *Journal-Chinese Ceramic Society*. 2000, 28:570–574.

White B, Banerjee S, O'Brien S, Turro NJ, Herman IP. Zeta-potential measurements of surfactant-wrapped individual single-walled carbon nanotubes. *The Journal of Physical Chemistry C*. 2007, 111:13684–13690.

Xu J, Feng W, Jiang L, et al. Influence of surfactants on chloride binding in cement paste. *Construction and Building Materials*. 2016, 125:369–374.

Yang Z, Gao Y, Mu S, et al. Improving the chloride binding capacity of cement paste by adding nano-Al2O3. *Construction and Building Materials*. 2019, 195:415–422.

Ye H, Jin X, Chen W, et al. Prediction of chloride binding isotherms for blended cements. *Computers and Concrete*. 2016, 17(5):655–672.

Yonezawa T. The mechanism of fixing Cl⁻ by cement hydrates resulting in the transformation of NaCl to NaOH. 8th International Conference on Alkali-Aggregate Reaction, 1989, 153–160.

Yuan Q. *Fundamental Studies on Test Methods for the Transport of Chloride Ions in Cementitious Materials*. Ghent University, Belgium; 2009.

Yue Y, Wang JJ, Basheer PM, Bai Y. Raman spectroscopic investigation of Friedel's salt. *Cement and Concrete Composites*. 2018, 86:306–314.

Zana R, Yeager EB. Ultrasonic vibration potentials. *Modern Aspects of Electrochemistry*, Springer; 1982, 1–60.

Zibara H, Hooton R, Thomas M, Stanish K. Influence of the C/S and C/A ratios of hydration products on the chloride ion binding capacity of lime-SF and lime-MK mixtures. *Cement and Concrete Research*. 2008, 38:422–426.

Zingg A, Winnefeld F, Holzer L, et al. Adsorption of polyelectrolytes and its influence on the rheology, zeta potential, and microstructure of various cement and hydrate phases. *Journal of Colloid and Interface Science*. 2008, 323:301–312.

Chapter 4

Chloride Binding and Its Effects on Characteristics of Cement-Based Materials

4.1 INTRODUCTION

Generally, the chloride ions within cement-based materials can be separated into two parts: free and bound chloride ions. Free state chloride ions refer to the soluble chloride ions in pore solution which can be migrated along with the transportation of pore solution. It is considered that the free chloride ions in pore solution dominate the chloride migration and chloride-induced corrosion of reinforced cement concrete. Within cement-based materials, chemical and physical reactions between chloride ions and cement hydrates can always occur, during which part of chloride ions in pore solution can be captured and slow down the chloride penetration process. Chloride ions being captured by cement hydrates are defined as bound chloride, in which physically adsorbed and chemically bound chloride ions are included. Due to the effects of chloride binding on chloride ingress process and chloride penetration resistance of cement-based materials, extensive studies have been conducted on effects of different factors on the chloride binding capacity of different cement-based and alkali-activated systems. In these studies, it is investigated that cement type and content, supplementary cementitious materials, water to binder (w/b) ratio, curing and exposure condition, and chloride source all have influences on the chloride binding capacities of cement-based materials. According to these studies, it has been confirmed that the contents of C_3A and C_4AF dominate the chemical binding of chloride ion, while C_3S and C_2S dominate the physical binding. Hydroxyl and sulfate ion may decrease the chloride binding capacity of cementitious materials.

The physical and chemical interactions between chloride ions and solid phase in cement-based materials can affect the process of chloride ingress. Chloride binding increases the total chloride content which can be captured by cement-based materials; on the other hand, the chloride penetration rate is reduced from chloride binding due to the decrease of free chloride concentration in pore solution. The formation of solid products from chloride binding can also change the structure and affect the properties of cement-based materials. This chapter discusses the chloride binding

77

of cement-based materials and its effects on chloride transport and micro-structure of cement-based materials.

4.2 CHLORIDE BINDING OF CEMENT-BASED MATERIALS

Usually, chloride binding occurs instantaneously or at least at a much greater rate than transport velocities. The pore system is always considered to be at equilibrium during the chloride binding process. This assumption may be valid, when the chloride ion travels slowly, just in the case of diffusion alone. However, this may not be valid when the ions are moving quickly and the test duration is short, as in the case of the rapid chloride migration test (RCM). In this case, transport would be occurring too quickly for equilibrium to be maintained (Barbarulo et al. 2000; Samson et al. 2003). Tang and Nilsson (1993) found that when crushed particles (with the size of 0.25–0.2 mm) were immersed in chloride solution, chemical binding was complete after less than 14 days. However, Arya et al. (1990) found that bound chlorides were still increasing after 84 days of immersion in 2% chloride solution. Olivier (2000) believed that the rate of chloride binding on crushed mortar particles is very high. Indeed, more than 80% of the bound chlorides are bound in less than 5 h.

Tang (1996) found that the rate of chloride binding changes the shape of chloride profile, although it does not remarkably change the penetration depth. Stanish (2002) carried out some studies to evaluate the binding rate of cement, and a high variation was observed in his research. However, the obtained solution has to be diluted before chloride titration to determine the chloride content of samples, which accounts for the high variation between results of different samples (Stanish 2002). According to the author, despite the variability, it was concluded that most binding occurred within the first hour or two of exposure.

The binding of chloride ion by cementitious materials is very complicated, and is influenced by many factors including chloride concentration, cement composition, hydroxyl concentration, cation of chloride salt, temperature, supplementary cementing materials, carbonation, sulfate ions, and electrical field, which has been concluded in Table 4.1. This section discusses how these factors affect the chloride binding in detail.

4.2.1 Chloride Concentration

Chloride concentration is probably one of the most important factors affecting chloride binding. Several studies have confirmed that a higher concentration of external chloride resulted in a higher chloride concentration in the pore solution and, consequently, a high level of chloride binding (Song et al. 2008a; Yuan et al. 2009; Machner et al. 2018). For a given type

Table 4.1 Factors Influencing Chloride Binding

Factors		Trend	Reasons	References
Chloride concentration↑		↑	More chloride ions can get to binding sites.	Tang and Nilsson (1993)
Cement	$C_3A + C_4AF$↑	↑	Formation of F salt and its analog.	Tang and Nilsson (1993), Suryavanshi et al. (1996)
	C_3S↑	Significant effect	Depends on Ca/Si ratio in C-S-H gel.	Beaudoin et al. (1990), Tang and Nilsson (1993)
	C_2S↑	↑		
	Sulfate↑	↓	The sulfate consumes C_3A and C_4AF, reducing the chemical binding.	Hassan (2001)
Mineral admixture	Silica fume↑	↓	Dilutes effect of C_3A on the chemical binding and formation of C-S-H with low Ca/Si ratio.	Page and Vennesland (1983)
	Fly ash ↑	↑	More Al_2O_3 form more F salt.	Dhir et al. (1997)
	Slag powder↑	↑	More Al_2O_3 form more F salt, or dilute sulfate content in cement.	Arya et al. (1990), Delagrave et al. (1997)
OH^- concentration↑		↓	Competition with Cl^- for binding sites and increases solubility of F salt.	Mehta (1977), Suryavanshi et al. (1996)
Cation of chloride salt		More binding from Ca^{2+} than from Na^+	Ca^{2+} binding of C-S-H surface produce more positive charge and thus more binding by the double electrical layer.	Hassan (2001), Suryavanshi et al. (1996), Wowra et al. (1997)
Temperature↑		↓	Higher temperature, more de-binding and larger solubility of F salt.	Hassan (2001)
Carbonation		↓	Decomposing of C-S-H and decreasing of porosity produce less binding sites.	Hassan (2001)
SO_4^{2-}↑		↓	SO_4^{2-} react with C_3A and its hydrated products to reduce formation of F salt.	Hassan (2001)

(Continued)

Table 4.1 (Continued) Factors Influencing Chloride Binding

Factors	Trend	Reasons	References
Electrical field	↓	Some physically bond chloride will be activated to de-binding	Ollivier et al. (1997), Shuguang et al. (2008)
Water-binder ratio↑	Significant effect between 0.3-0.5	Depends on porosity and hydration degree.	Hassa (2001)
Curing age↑	Binding amount↑, Binding capacity↓	The binding gradually gets saturated.	Yu et al. (2007)

of cement, a maximum chloride binding capacity exists. Under this limit, the higher the chloride concentration in the pore solution is, the more chloride ions have the chance to access to the binding sites, and the higher the chloride binding will be (Yang et al. 2019). The relationship between free chloride and bound chloride is known as binding isotherm, which will be discussed in detail in Section 4.3.

4.2.2 Cement Composition

4.2.2.1 C_3A and C_4AF

In the case of internal chloride, Rscheeduzzafar (1990) found a substantial reduction in water-soluble chloride content with increased C_3A content. In addition, they also found X-ray diffractogram evidence of calcium chloroaluminate formation in chloride-penetrated concrete made with cement with 9% and 14% C_3A. Blunk et al. (1986) noticed that a pure C_3A-gypsum mixture bound more chlorides than an ordinary Portland cement (OPC) and C_3S paste when they were treated with chloride solutions of varying concentrations. Arya et al. (1990) stored OPC and sulfate resistant Portland cement (SRPC) in a 20 g/l NaCl solution, and found that SRPC bound considerably less chloride than OPC, because of a lower C_3A content. Several studies have been carried out in the case of internal chloride to study the influence of C_3A on chloride binding (Delagrave et al. 1997; Kim et al. 2016; Vu et al. 2017; Ann and Hong 2018); all the results indicated that higher C_3A content resulted in higher binding capacity. Glass and Buenfeld (2000) developed a model to predict the free chloride concentration as a function of various input variables. The effect of C_3A content on the binding isotherm is shown in Figure 4.1. It can be seen that the chloride binding increases significantly as the C_3A content increases.

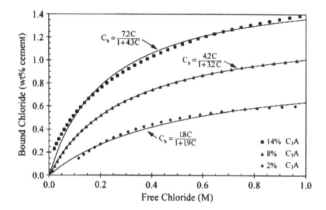

Figure 4.1 Predicted binding data together with the fitted Langmuir adsorption isotherms for a 0.45 w/c OPC cement paste. (From Glass and Buenfeld 2000.)

Due to the reactions between chloride ion and C_3A or C_4AF, which result in the formation of Friedel's salt and its analogs, the contents of C_3A and C_4AF in cement dominate chemical binding. The formation of Friedel's salt may be due to direct reaction between C_3A and $CaCl_2$, according to the following reaction schemes when NaCl is present (Hassan 2001; Qiao et al. 2018):

$$Ca(OH)_2 + 2NaCl = CaCl_2 + 2Na^+ + 2OH^- \tag{4.1}$$

$$C_3A + CaCl_2 + 10H_2O \rightarrow C_3A \cdot CaCl_2 \cdot 10H_2O \tag{4.2}$$

According to Suryavanshi et al. (1996), Friedel's salt and its analog forms by two separate mechanisms: adsorption and anion-exchange. In the adsorption mechanism, the formation of Friedel's salt is due to the adsorption of bulk chloride ion present in the pore solution into the interlayers of the principal layers, $[Ca_2Al(OH)_6 \cdot 2H_2O]^+$, of the aluminate ferrite mono (AFm) structure to balance the charge. In the anion-exchange mechanism, a fraction of the chloride ion replaces OH^- ions present in the interlayers of the principal layer of the AFm hydrates (C_4AH_{13} and its derivations) to form Friedel's salt, which is expressed as follows:

$$R - OH^- + Na^+ + Cl^- \rightarrow R - Cl^- + Na^+ + OH^- \tag{4.3}$$

where R is the principle layer of the AFm hydrates of composition, $[Ca_2Al(OH)_6 \cdot nH_2O]^+$.

As a result of Friedel's salt formation by adsorption mechanism, some Na^+ ions, equivalent to the adsorbed chloride ions, are removed from the pore solution to maintain the ionic charge neutrality. In contrast, the Friedel's salt formation by the anion-exchange mechanism involves the release of OH⁻ ions from the AFm hydrates into the pore solution, thereby increasing the pH of the pore solution. It is generally accepted that the amount of bound chlorides increases with increased alumina-bearing phase. Under the action of carbonation or intruded sulfate, the chemical binding is reversible (Saillio et al. 2014; Chang 2017).

4.2.2.2 C_3S and C_2S

Compared to C_3A, fewer studies were focused on C_3S and C_2S. Calcium-silicate-hydrate (C-S-H) gel is the main hydration product of Portland cement and dominates the physical binding of chloride. Physical binding is due to the adsorption of chloride ion to the C-S-H. Ramachandran (1971) distinguished three types of interaction with C-S-H: Chlorides can either be present in a chemisorbed layer on the hydrated calcium silicates, penetrate into the C-S-H interlayer spaces, or be intimately bound in the C-S-H lattice. The higher the content of C_3S and C_2S is, the higher physical binding of chloride will be. Tang and Nilsson (1993) found that the chloride binding capacity of OPC concrete is strongly dependent on the C-S-H content in the concrete, regardless of the water–cement ratio and the addition of aggregate. Beaudoin et al. (1990) suggested that the binding capacity of C-S-H depends on its calcium/silica (C/S) ratio. Lower C/S ratios result in lower binding capacities. However, by adding chloride to pure alite paste, Lambert et al. (1985) found that C-S-H only absorb a very small amount of chloride ions.

While C_3A and C_4AF dominate the chemical binding of chloride ion, C_3S and C_2S dominate physical binding. Hassan (2001) treated pure C_3A, C_4AF, C_3S, and C_2S, respectively, with chloride solutions. He found that C_3A plays the most important role in the chloride binding capacity of cement. C_3A is a good indicator of binding capacity of cement in the high concentration range (1.0–3.0 mol/L), but it is not a good indicator of chloride binding at the low concentration (0.1 mol/L). The chloride binding capacity of C_4AF is one third of that of C_3A. The contribution of C_3S to the binding capacity could possibly range from 25% to over 50%.

4.2.2.3 SO_3 Content

The SO_3 content of cement also affects the chloride binding. Hassan (2001) found that SO_3 had a negative influence on the binding capacity, especially at low chloride concentrations (0.1 mol/L). Sulfates can react with C_3A and C_4AF to form ettringite or monosulfate. As the chloride concentration increases, the monosulfate first transforms into Kuzel's salt, and then into Friedel's salt at higher concentrations. Ettringite also starts to transform at higher concentration (3.0 mol/L). This could explain why the influence is more significant at low concentration.

4.2.3 Supplementary Cementitious Materials

At present, supplementary cementitious materials (SCMs) such as silica fume, slag, and fly ash are widely used as cement replacements for material cost, environmental reasons, or improvement of the properties of concrete. Each cementitious material has different chemical composition and physical properties, which results in different effects on chloride binding capacity.

4.2.3.1 Fly Ash

Thomas et al. (2012) investigated the effects of SCMs on chloride binding in hardened cement paste; the results showed that fly ash replacement (25%) showed higher chloride binding capacity from higher alumina contents and the subsequent formation of C-A-S-H. The increasing physical adsorption on the surface of hydration or pozzolanic reaction products, such as C-S-H, C-A-H, ettringite, and monosulfate were also considered as the enhancement of chloride binding capacity of fly ash-blended cement.

Cheewaket et al. (2010) studied the effects of fly ash on the chloride binding capacity of cement paste; the results showed that as the replacement level of fly ash increased to 50%, the chloride binding capacity of samples gradually increased. By using the equilibrium method, Dhir et al. (1997) found that the chloride binding capacity of cement paste increases with the increase in fly ash replacement level up to 50%, and then declines at 67%. Many other researchers (Arya et al. 1990; Byfors 1986; Wiens 1997) also found that partial replacement of cement with fly ash had a positive effect on the chloride binding of the cement paste exposed to chloride environment. However, Nagataki et al. (1993) found that the replacement of 30% cement with fly ash reduced the chloride binding capacity of cementitious material in the case of external chlorides. There is no good explanation for this. The decrease of chemical binding capacity with class F fly ash replacement was also found by Azad and Isgor (2016), which was explained by two main factors: (1) the larger available alkali content in class F fly ash than in OPC cement; and (2) the lower C3A content of fly ash-blended cement. In the case of internal chloride, the increase in chloride binding due to the replacement of fly ash was found (Arya et al. 1990; Arya and Xu 1995; Byfors 1986). An increase in chloride binding may be mainly ascribed to the high alumina content in fly ash, which results in the formation of more Friedel's salt (Wiens 1997).

4.2.3.2 Ground Granulated Blast Furnace Slag (GGBFS)

Many researchers (Arya et al. 1990; Hassan 2001; Nagataki et al. 1993; Khan and Kayali 2016; Kopecskó and Balázs 2017) have shown that partial substitution of cement with GGBFS increases the chloride binding in the case of external chlorides. For GGBFS-blended cement, the phases able to bind chloride are mainly AFm and calcium aluminum silicate hydrate (C-A-S-H) (Florea and Brouwers 2014). Kayali et al. (2012) reported that the chloride binding of GGBFS came from two mechanisms. The first is the formation of hydrotalcite during hydration due to high content of magnesia in GGBFS; the second is the formation of Friedel's salt from the aluminum phase in GGBFS. The X-ray diffractometry (XRD) analysis revealed that the content of hydrotalcite formed in hardened pure GGBFS pastes reached 54% of total crystallized phase, higher than that of C-S-H gel in Portland cement paste (around 40%). It was concluded in their paper that hydrotalcite was responsible for superior chloride binding capacity of GGBFS-blended concrete. This conclusion was further confirmed by Khan et al. (2016), who also reported that the ability of hydrotalcite to bind chlorides was not significantly impaired by the competitive adsorption of carbonates. However, a recent paper (Maes et al. 2013) obtained totally different results wherein the chloride binding capacity decreased when GGBFS was added as cement replacement. According to the discussion in this paper, Al_2O_3 and Fe_2O_3 played similar roles in chloride binding, and the lower content of Fe_2O_3 for the GGBFS used in their study was responsible for the lower chloride binding capacity of GGBFS-blended concrete. Sun et al. (2010) analyzed the chloride binding in the cement-GGBFS paste by XRD, differential scanning calorimetry (DSC) and thermogravimetry-differential thermogravimetry (TG-DTG). The results showed that the chloride binding capacity of GGBFS was closely related to the specific surface area and chemical composition of GGBFS; it was enhanced with higher specific surface area and lower sulfur trioxide content.

The increase in chloride binding due to the replacement of GGBFS was also observed in the case of internal chloride (Arya et al. 1990; Arya and Xu 1995; Potgieter et al. 2011). Yu et al. (2007) determined the equilibrium chloride ion concentration of concrete containing added sodium chloride with different concentrations by isothermal adsorption method and calculated the total, physical, and chemical binding capacity of GGBFS-blended concrete to chloride ions. The results showed that with the increase of GGBFS content in samples, the total and chemical chloride binding capacity of concrete samples ascend in first and descend at last, and samples with 40% (mass percentage to total cementitious materials) represented the higher chloride binding capacity, while no obvious effects were found on physical adsorption capacity of samples with different GGBFS content. Dhir et al. (1997) suggested that high binding capacity of cement-GGBFS paste may be due to the high alumina content in GGBFS, resulting in the acceleration of Friedel's salt formation. Arya et al. (1990) suggested that the increase in adsorbed chlorides was responsible for the higher level of binding. It is worthwhile to mention that Xu (1997) studied the chloride binding of cement paste with admixed chloride and found that when the sulfate ion in GGBFS-cement paste was raised to the same level as cement paste, the higher binding capacity disappeared. Therefore, he ascribed the higher binding capacity of GGBFS-cement to the dilution effects of sulfate ions.

4.2.3.3 Silica Fume

Different from GGBFS and fly ash, the incorporation of silica fume in cement-based materials is generally regarded to lower the binding capacity (Thomas et al. 2012; Jung et al. 2018). According to the studies of Nilsson et al. (1996), the addition of silica fume influenced binding in three ways: (a) dilution of the C_3A which may reduce chemical binding, (b) reduction of the pH value of the pore solution which should increase chloride binding, and (c) an increase in the amount of C-S-H which may increase physical binding of chlorides. Besides, the addition of silica fume also reduces the C/S ratio, which can result in lower physical adsorption capacity of C-S-H (Koleva et al. 2007). Figure 4.2 shows the chloride binding isotherms of concrete with different silica fume replacement levels after five years of exposure to tidal zone (Dousti et al. 2011). The use of silica fume significantly decreased the bound chloride content of concrete, while the influences of replacement level were not that noteworthy.

Arya et al. (1990) studied the chloride binding behaviour of cement paste with 15% silica fume by exposing the cement paste to 0.56 mol/L NaCl solution, and found that partial replacement of cement with silica fume decreased chloride binding. Some researchers observed a different trend (Byfors 1986). However, most studies (Hassan 2001; Thomas et al. 2012) showed that silica fume decreased chloride binding in the case of external chloride. In the case of internal chloride, many studies (Arya et al. 1990; Arya and Xu 1995; Hussain and Al-Gahtani 1991) observed the same phenomenon as external chloride. Three changes may be induced by partial replacement of cement with silica fume: (1) an increase of the content of C-S-H with lower C/S ratio; (2) a reduction in pH value; (3) a dilution effect of C_3A. The increase in C-S-H may be in favor of binding, but, the lower C/S ratio (Beaudoin et al. 1990) may have a negative impact on binding. Page and Vennesland (1983) found that the amount of Friedel's salt decreased with increasing silica fume content by using differential thermal analysis and thermogravimetry. The reduction in pH may be responsible for it. The dilution effect of C_3A may decrease the binding as well.

Generally, the amount of bound chloride in SCMs-blended cement strongly depends on the quantities of alumina in the binders, especially for chemical bound chloride. Therefore, we can easily recognize the chloride binding capacity of different SCMs basically. The high content of Al_2O_3 in metakaolin (Thomas et al. 2012) gave totally different effects on chloride binding capacity of binders to that of limestone with low content of alumina (Ipavec et al. 2013). Besides the alumina content, it was also presented that the calcium to alumina (C/A, C-A-S-H) and C/S (C-S-H) ratios also played a role in chloride binding capacity of SCMs. Higher C/A ratio of C-A-S-H

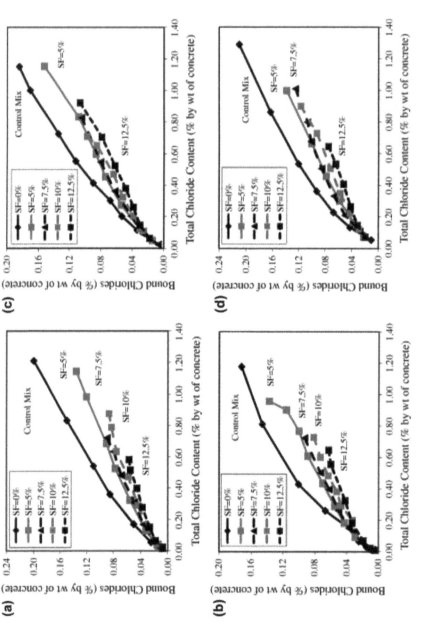

Figure 4.2 Chloride binding isotherms of concrete samples made of different silica fume replacement level after 5 years for (a) w/b = 0.35, (b) w/b = 0.40, (c) w/b = 0.45, and (d) w/b = 0.50. (From Dousti et al. 2011.)

and C/S ratio of C-S-H show greater binding capacity (Zibara et al. 2008). Saillio et al. (2015) reported that the physical adsorbed chloride increased with the SCM (fly ash, GGBFS or metokalin) content of the cement paste due to the different types of C-S-H produced during hydration in Portland cement and SCM cement pastes.

4.2.4 Hydroxyl Ion Concentration

Tritthart (1989) immersed cement pastes in chloride solutions with different pH values, and found that chloride binding increased with decreased pH value. Many researchers (Page et al. 1991; Sandberg and Larsson 1993) also found that the hydroxyl concentration in the external environment has a significant influence on chloride binding. The general tendency is that the higher the hydroxyl concentration is, the lower the amount of bound chloride is. As pointed out by Tritthart (1989), a competition exists between hydroxyl and chloride ion for adsorption sites on cement surface. Suryavanshi et al. (1996) also suggested that a competition exists between hydroxyl and chloride ion during their adsorption in the interlayers of the principal layers, $[Ca_2Al(OH)_6 \cdot 2H_2O]^+$. Roberts (1962) suggested that an increased pH of the chloride solution increased the solubility of Friedel's salt, thereby releasing chloride to the pore solution and reducing the amount of chemically bound chloride. Compared to other factors, the effects of hydroxyl ion concentration on chloride binding capacity of cement-based materials is relatively small. In study of Song et al. (2008b), no direct relationship between pH value of pore solution and chloride binding capacity of paste was obtained.

4.2.5 Cation of Chloride Salt

Several studies (Arya et al. 1990; Blunk et al. 1986; Delagrave et al. 1997; Wowra et al. 1997) have noticed that cations of chloride salts affect chloride binding. Delagrave et al. (Delagrave et al. 1997) found that $CaCl_2$ resulted in more bound chlorides than NaCl. Arya et al. (1990) found that NaCl resulted in 43% bound chloride, $CaCl_2$ 65%, $MgCl_2$ 61%, and seawater 43%. The nature of the associated cation probably has a predominant influence on the chloroaluminate solubility and the accessibility of chloride ion to the adsorption sites (Delagrave et al. 1997). In addition, the presence of Na^+ in hardened concrete results in a higher pH value than that of Ca^{2+} and Mg^{2+}. Thus, the degree of competition offered by OH^- in NaCl solution is higher than that in $CaCl_2$ and $MgCl_2$ solution. Zhu et al. (2012) studied the bound chloride content of concrete with different w/b ratios. As shown in Figure 4.3, the chloride binding capacity was influenced by the tendency of bound chloride is influenced by cation of chloride salt; the tendency was C_b ($CaCl_2$) > C_b ($MgCl_2$) > C_b (KCl) ≈ C_b (NaCl). Wowra et al. (1997)

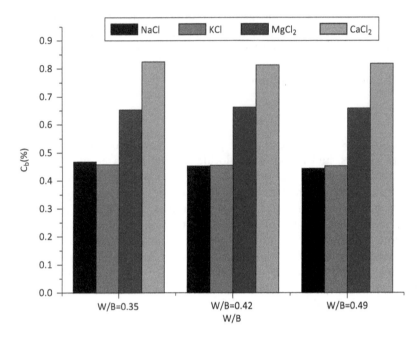

Figure 4.3 Effects of cation ions of chloride salt on bound chloride content of concrete. (From Zhu et al. 2012.)

suggested that an increase in binding was induced by the adsorption of Ca^{2+} ions on the surface of C-S-H, which results in an increased positive charge and a subsequent higher adsorption of chloride ion in the double electrical layers.

4.2.6 Temperature

Larsson (1995) and Roberts (1962) found that the amount of bound chloride decreased as temperature increases. Indeed, for physical adsorption, an elevated temperature increases the thermal vibration of absorbates, resulting in more unbound chloride (Jensen et al. 2000). For chemical reaction, although an elevated temperature increases the reaction rate, it may also increase the solubility of the reaction products (Friedel's salt), resulting in more reactants free at the equilibrium. Hassan (2001) found that at a low chloride concentration (0.1 and 1.0 mol/L), an increased temperature resulted in a decreased binding, while at high chloride concentration (3.0 mol/L), an increased temperature resulted in an increased binding. It was also found that in the presence of limestone, the low temperature (5°C) suppressed the chloride binding capacity of cements (Ipavec et al. 2013).

4.2.7 Carbonation

There are few publications dealing with the effect of carbonation on chloride binding. Hassan (2001) immersed three different pre-carbonated pastes into chloride solutions with different concentrations and found that the pastes had almost no binding capacity. It was found that within cement-based materials, the chloride ions potentially ingressed from carbonated to un-carbonated zone (Ye et al. 2016).

Carbonation changes the nature of hydration products and will definitely have a great effect on the chloride ion binding of the cement pastes. After carbonation, the hydration products are converted to $CaCO_3$, silica gel and alumina gel, and the pH value of the system drops to about 9. For physical binding, the decomposition of C-S-H and the reduction in total porosity may provide fewer sites for ion exchange reaction and physical binding (Liu et al. 2016a). For chemical binding, the decrease of pH value due to carbonation may decrease the degree of competition offered by hydroxyl but increase the solubility of Friedel's salt (Suryavanshi and Swamy 1996). Overall, carbonation decreases the chemical binding capacity of cement-based materials. This was verified by the XRD and DTA results from Suryavanshi and Swamy (1996); they found that under the action of carbonation, chlorides from Friedel's salt were released into pore solution. Thus, in the presence of carbonation, reinforcement concrete was at higher corrosion risk. Saillio et al. (2014) investigated the chloride binding isotherms of uncarbonated and carbonated cementitious materials by the equilibrium method; a decrease of chemical bound chloride (Fridel's salt) and C-S-H bound chloride was obtained. For the former, lack of portlandite and modification of aluminate phases equilibrium were considered as the reason, and the modification of surface charge of C-S-H gel during carbonation explained the decrease of C-S-H bound chloride. It was found (Liu et al. 2017) that the carbonation reaction significantly decreased the bound chloride in concrete and in carbonation zone; for some specimens, the free chloride content almost equalled to the total chloride content (Liu et al. 2016b). Based on the retardation of carbonation on chloride binding and chloride migration in cement concrete, Zhu et al. (2016) proposed a comprehensive model to study the combined carbonation and chloride ingress in concrete.

4.2.8 Sulfate Ion

It is well known that a sulfate ion can react with C_3A during its hydration process to form monosulfoaluminate or ettringite (Cao et al. 2019). Byfors (1986) measured chloride penetrated into cement pastes after immersion in 0.28 mol/L NaCl solutions containing sulfate ions for eight months, and found that increased sulfate ion concentrations resulted in a slight decrease of chloride binding. It is generally considered that the presence of sulfates decreases the chloride binding of hardened cement paste. Wowra et al. (1997), Frias et al. (2013), and Sotiriadis et al. (2017) observed similar phenomena. Sulfate ions preferentially react with C_3A and its hydration products, which results in the decomposition of Friedel's salt (Geng et al. 2015; De Weerdt et al. 2014; Xu et al. 2013). Brown and Badger (2000) investigated the distributions of bound sulfates and chlorides in concrete subjected to mixed NaCl, $MgSO_4$, and Na_2SO_4 attack, and found that the microstructure of these concretes were characterized by zones of gypsum near the top and bottom surfaces, zones of ettringite adjacent to these, and central zones of monosulfoaluminate and Friedel's salt. The formation of Friedel's salt was due to the higher penetration velocity of chloride ion than sulfate ion. Thus, sulfate ions may decrease the chloride binding, whether sulfate ions were added into concrete or penetrate into concrete from an external environment. Besides, it was found that the addition of $MgSO_4$ decreased the pH value of pore solution. Therefore, the surface charge of C-S-H gel and physical adsorption chloride content decreased (Tran et al. 2018).

4.2.9 Electrical Field

Natural diffusion test and rapid chloride migration test are widely applied in evaluating the chloride penetration resistance of cement-based materials. Especially for the latter, the applied electrical field can markedly reduce the testing period. However, the applied external field potential affects the binding capacity and desorption process of cement-based materials.

Applying the external voltage shortens the duration of the test at the cost of possible changes in the pore structure and the number of chemically bound chloride ions (Zheng et al. 2016). At the same time, the external electric field would affect the charge distribution on the surface of the hydration products and the electrical double layer at the solid–liquid interface (Liu and Shi 2009). Therefore, the use of external voltage will have some unexpected effects on the chloride penetration and binding capacity. Several researches have been carried out to investigate the effects of applied voltage on chloride ion binding and to compare results with results obtained by diffusion method (Ma et al. 2013; Spiesz and Brouwers 2012, 2013).

Some researchers (Krishnakumark 2014; Spiesz and Brouwers 2013; Voinitchi et al. 2008) reported that the adsorption of chloride ions after reaching the steady state was independent of the applied voltage, and the total chloride content was increased with the increase of chloride concentration in the external environment. Also, Yuan (2009) obtained similar chloride adsorption isotherm for samples after natural diffusion and RCM tests under the same curing conditions and ages. Furthermore, a study by Ollivier et al. (1997) showed that the external applied field within the voltage range of 2~30 V had no obvious impact on chloride binding capacity of cement mortars. Numerical simulation on diffusion and RCM tests at the non-steady state showed that the free chloride ions in pore solution could be instantaneously combined and no variation of free chloride content was found between samples after diffusion and migration test (Spiesz and Brouwers 2012).

Spiesz and Brouwers (2013) showed that it took seven days or more for chloride binding in cement matrix to reach an equilibrium in diffusion method, which was much longer than the testing duration of RCM test. Castellote et al. (1999, 2001) analyzed the chemically bound chloride and free chloride contents of samples after RCM test based on X-ray fluorescence technique and leaching method, respectively. In their studies, the obtained chloride adsorption isotherm was compared with the diffusion chloride adsorption isotherm from Sergi et al. (1992). The applied electric field suppressed chloride binding at lower free chloride concentration (<97 g/L) while enhanced it at higher concentrations. The decreased contact time and altered double layer potential of the pore walls were considered as influential causes. Also, Gardner (2006) observed that the implementation of RCM test for two weeks prior to the diffusion method of more than 180 d reduced the amount of chemically bound chloride in cement mortar by up to 50%. The external electric field permanently changed the chloride binding capacity of the matrix. However, in the above-mentioned studies, only chloride adsorption isotherm of samples was considered, and the physical adsorption chloride on the surface of hydration products was ignored. Besides, the differences between testing duration in diffusion and rapid migration was neglected in previous studies in order to ensure the same soaking time in the chloride solution. Based on this, effects of applied voltage on free state chloride ions and chloride binding—physically adsorbed and chemically bound—in pore solution and electrical double layer need further attention.

In our studies, different types of chloride ions of cement pastes after natural diffusion (Group A) and RCM (Group B) tests have been studied. As shown in Table 4.2, the applied voltage mainly affected the water-soluble chloride, including physically bound chloride in EDL and free chloride ions in bulk pore solution obtained by pore solution expression, and had no obvious effect on the content of Friedel's salt. This signified that the reaction between chloride ions and C_3A can be completed in a relatively short time, and the applied voltage brought no effects on this reaction. However, no Friedel's salt was detected in samples soaked in the solution with 0.1 mol/L NaCl. This can be explained by the formation mechanism of Friedel's salt. The penetrated chloride ions may react with aluminum phase in hardened cement and form Friedel's salts as the chloride concentration increased. The stability of Friedel's salt is closely related to the chloride concentration in the surrounding environment and the salt crystals may decompose when chloride concentration decreases. In the diffusion test at low concentration, the diffusion rate of chloride ions was slow and

Table 4.2 Effect of Electrical Field on Different Types of Chloride Ions

Sample	Total Chloride (%)		Water-Soluble Chloride (%)[a]		Chemically Bound Chloride (%)		C-S-H Bound Chloride (%)	
	Group A	Group B	Group A	Group B	Group A	Group B	Group A	Group B
PC	1.48	1.86	0.34	0.57	0.56	0.55	0.57	0.74
20% SL	1.52	1.90	0.44	0.60	0.56	0.57	0.52	0.73
40% SL	1.63	1.96	0.43	0.65	0.68	0.68	0.52	0.63
60% SL	1.37	1.75	0.42	0.64	0.86	0.85	0.10	0.26

[a] The chloride concentration in expressed pore solution is transferred to chloride mass content based on MIP test results.

the chloride concentration in the pore solution was too low to form stable Friedel's salt. However, the external voltage could sharply increase the chloride concentration in the pore solution within a short period of time. The results show that the applied voltage can also increase the free chloride concentration in pore solution. According to the results, the applied voltage had no effects on chemical binding of cement paste when the chloride concentration in pore solution exceeded a specific value. The study of Xia and Li (2013) also showed that when the external voltage was applied, the migration of chloride ion in cement-based materials was determined by the initial chloride concentration in the pore solution.

4.3 CHLORIDE BINDING ISOTHERM

The relationships between free and bound chloride ions over a range of chloride concentrations at a given temperature are known as the chloride binding isotherms. Until now, four types of binding isotherm (i.e., linear, Langmuir, Freundlich and BET binding isotherm, Table 4.3) have been proposed to describe the relationship, which are described in the following sections.

4.3.1 Linear Binding Isotherm

Tuutti (1982) proposed a linear binding isotherm, which can be expressed as follows:

$$C_b = \kappa C_f \qquad (4.4)$$

where C_b is the concentration of bound chlorides, k is a constant and C_f is the concentration of free chlorides. It is valid for free chloride concentrations lower than 20 g/l. Arya et al. (1990) proposed a linear relationship

Table 4.3 The Four Binding Isotherm Used in Publications

Isotherm	Range	Remark	References
Linear	Valid in concrete immersed into chloride salt solution	Leaching of OH- produce a the linear relationship.	Hassan (2001), Mohammed and Hamada (2003)
Langmuir	Low Cl- concentration (<0.05 mol/L)	It means that all binding sites are occupied in the case of high Cl- concentration.	Tang (1996), Tang and Nilsson (1993)
Freundlich	High Cl- concentration (>0.05 mol/L)	It includes two important order of chloride concentration in seawater.	Tang (1996), Tang and Nilsson (1993)
BET	Cl- concentration <1.0 mol/L		Xu (1990)

with an intercept on the axis for bound or total chlorides. Ramachandran et al. (1984) found that the linear binding isotherm was not valid based on their results. The non-linear relationship between free and total chlorides is now generally accepted. The linear relationship is an oversimplification and seems to be applicable within a limited range of chloride concentrations (Olivier 2000). It overestimates chloride binding at high chloride concentration, and underestimates chloride binding at low chloride concentrations, as shown in Figure 4.4.

However, some researchers (Mohammed and Hamada 2003; Sandberg 1999) found that the relationships between free chloride and total chloride contents in field-exposed concrete were linear, and the linear chloride binding isotherm was also applied in some chloride penetration models (Oh and Jang 2007). Mohammed and Hamada (2003) found that there were linear relationships between free chloride and bound chloride in concrete based on several long-term (ranging from 10 years to 30 years) exposure tests under marine environment for various types of cement, such as ordinary Portland cement, high early strength Portland cement, moderate heat Portland cement, calcium aluminate cement, slag cement, and fly ash cement. It should be noted that the water-soluble chloride concentration is defined as free chloride according to Mohammed and Hamada (2003), and the acid-soluble chloride concentration is defined as total chloride. In another study, Sandberg (1999) found a linear relationship between free and total chloride concentration in field-exposed concrete, which was different from the corresponding relationship measured in equilibrium experiments. He ascribed the difference to the leaching of hydroxyl. In his study, the free chloride concentration was obtained by the pore solution expression method, and the acid-soluble chloride was defined as the total chloride.

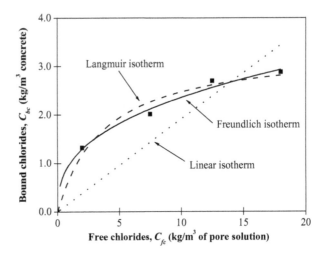

Figure 4.4 Plots of the Linear, Langmuir and Freundlich isotherms. (From Hassan 2001.)

4.3.2 Langmuir Isotherm

The Langmuir isotherm, derived from physical chemistry, is assuming monolayer adsorption, which explains that the slope of the isotherm curve at high concentrations approaches zero. It is of the following form:

$$C_b = \alpha C_f / (1 + \beta C_f) \tag{4.5}$$

where a and b are constants, which vary with the binder composition. These coefficients are obtained by non-linear curve-fitting of the experimental data and have no physical meanings. In the case of chloride binding, the Langmuir isotherm indicates that all the absorption sites are occupied by chloride ion at high free chloride concentration (Papadakis 2000). Sergi et al. (1992) used the Langmuir isotherm to account for the relationship between free and bound chloride, and obtained values of *a* and *b* as 1.67 and 4.08, respectively, by linear regression analysis of data from OPC paste samples with w/c = 0.5 (C_f and C_b were expressed by the authors as mol/ l and mmol/g of cement, respectively). Obviously, different units of free and total chloride content used in the isotherm can result in different *a* and *b*. Tang and Nilsson (1993) found that the Langmuir isotherm was an excellent fit of their binding data at concentrations lower than 0.05 mol/L.

4.3.3 Freundlich Binding Isotherm

The Freundlich binding isotherm can be expressed as follows:

$$C_b = \alpha C^\beta_f \tag{4.6}$$

where a and b are binding constants. Tang and Nilsson (1993) suggested that monolayer adsorption occurred at low concentrations (which was better described by the Langmuir isotherm), but that adsorption became more complex at concentrations higher than 0.05 mol/L and was described better by the Freundlich isotherm. The difference between the Freundlich and Langmuir isotherms is their behavior at high concentrations, as shown in Figure 4.4. It was found by Tang and Nilsson (1993) that the Freundlich equation fitted the data very well in a range of free chloride concentrations from 0.01 to 1 mol/L. This range covered the two most important magnitude orders of free chloride concentration in sea water. Both Olivier (2000) and Tang (1996) used Freundlich binding isotherm in their models to predict the chloride transport in concrete. By using Freundlich binding isotherm, Weiss et al. (2018) proposed an approach to predict chloride ingress in saturated concrete, and the simulation results corresponded well with the experimental results.

4.3.4 Brunauer, Emmett, Teller (BET) Isotherm

Brunauer, Emmett, Teller (BET) is originally applied to gas adsorption. Xu (1990) used a modified BET isotherm to describe the chloride binding:

$$
\frac{c_b}{c_{bm}} = \frac{\alpha \frac{c}{c_s}\left[1-(1-\beta)\left(1-\beta\frac{c}{c_s}\right)^2\right]}{\beta\left(1-\beta\frac{c}{c_s}\right)\left(\beta\frac{c}{c_s}+\alpha\frac{c}{c_s}\left(1-\beta\frac{c}{c_s}+\frac{c}{c_s}\right)\right)}
\tag{4.7}
$$

The equation was also used by Tang (1996). It was found that BET equation corresponded very well with the experimental results when the free chloride concentrations were lower than 1.0 mol/L. However, no report could be identified how it fits experimental results when free chloride concentration is greater than 1.0 mol/L.

4.4 EXPERIMENTAL DETERMINATION OF BINDING ISOTHERM

The coefficients used in isotherm equations are usually determined by various experiments. This section reviews the common methods used to determine binding isotherms, including equilibrium method, pore expression method, diffusion cell method, and migration cell method.

4.4.1 Equilibrium Method

The most straightforward and widely applied test method is the equilibrium method, which is considered to be quite accurate. In the equilibrium method, samples are simply put in a known chloride solution until equilibrium is reached. However, reaching equilibrium takes a quite long time. Up to one year is required for cement pastes which were 10 mm thick (Tritthart 1989). To shorten the time to reach equilibrium, Tang and Nilsson (1993) immersed a few grams of particles in a salt solution having a certain chloride concentration. The content of bound chloride is obtained by calculating the difference between the concentration of the initial solution and the concentration after a certain period of exposure, when the equilibrium is reached. According to Tang and Nilsson (1993), with the particle size of 0.25–2 mm, the adsorption equilibrium could be reached within 14 days. After equilibrium, the inside solution was pipetted to determine the chloride concentration by potentiometric titration using 0.01 mol/L $AgNO_3$ and selective electrode. The chloride concentration of the inside solution can also be measured by using thin-film X-ray fluorescence spectrometry on salts precipitated by evaporation from the solutions chloride ion (Dhir et al. 1995). By immersing the particles in a series of solutions with different concentrations, the complete binding isotherm can be obtained.

There are some concerns on this method. First, it does not take into account the effect of leaching of other species. In fact, many other ionic species exist in the pore solution, especially hydroxyl ions, which affect binding. Secondly, to shorten the time to reach equilibrium, the sample is required to be crushed to the size of 0.25–2 mm, which may result in carbonation and allow further hydration. This may affect the test accuracy (Glass et al. 1996). Actually, it is very difficult to reach equilibrium. Arya et al. (1990) found that bound and free chloride ions were still increasing after 84 days of immersion in 2% chloride solution.

4.4.2 Pore Solution Expression

In this method, pore solution is expressed by pressing the concrete sample under very high pressure. The pore solution is then analyzed by chemical procedures in order to obtain the free chloride concentration (Larsen 1998). This method needs special equipment. Sometimes, it is difficult to obtain enough pore solution of concrete, especially in the case of a low water to binder ratio. Sergi et al. (1992) used this method to obtain a Langmuir adsorption isotherm.

The method can avoid some potential problems, such as carbonation, associated with the use of small samples required by other methods. Furthermore, a complete binding isotherm may be obtained on a single specimen. However, some loosely bound chloride ions may be released into the solution under pressure. This results in overestimating the free chloride ion concentration. The chloride concentration could be 20% higher than the true value (Glass et al. 1996).

In our recent studies, as shown in Chapter 3, the effects of pore solution expression on microstructure and chloride content of cement pastes were examined. During the process of pore solution expression, the high external applied pressure compressed the cement paste and made the paste more compact. The increase of compactness extracted the pore solution inside and re-arranged the inner pore structure. According to the NMR measurement results, the peaks in NMR signal plots with the T2 relaxation time larger than 0.4 ms (or pore size larger than 40 nm) are almost erased. Therefore, the pore solutions in pores with large diameter were totally extracted by pore solution expression. For small pores with diameter smaller than 40 nm, the peak value of the signal amplitude of NMR measurement was significantly decreased. It has been investigated in previous studies (He 2010; He et al. 2016) that the pore solution expression experiment decreased the total porosity and percentage of large pore in total porosity of cement pastes, which was highly consistent and confirmed by the results obtained here.

As a widely applied method to obtain and study the pore solution of cement-based materials, more studies are still needed to answer which part of the pore solution within pore can be extracted, and how does the pore structure of cement paste look after the process of pore solution expression? Studies are required to provide theoretical foundation for the application of pore solution expression method in studying the pore solution of cement-based materials and chloride binding.

4.4.3 Diffusion Cell Method

Glass et al. (1998) proposed that the chloride binding isotherm could be obtained from diffusion cell experiments used to determine the steady state diffusion coefficients. After the steady state diffusion experiment, the specimens were removed from the cells and then frozen in nitrogen and subsequently ground to powder samples. The acid-soluble chloride content of these samples and the chloride content of the storage solutions were determined by potentiometric titration against silver nitrate. The free chloride concentration is estimated by assuming that the concentration gradient is linear under steady state diffusion conditions (i.e., the free chloride concentration in specimens does not change with time). Glass's experimental results showed that the chloride binding obtained from this method is typical as those obtained by other methods, such as equilibrium method and pore solution expression. Bigas (1994) obtained binding isotherm from a series of diffusion tests with different upstream chloride concentrations. In the diffusion test, the steady state flow and the time lag is determined for each specimen. The full binding isotherm can be determined from the time lag as a function of chloride concentration. Diffusion cell method is time consuming and not easy to manipulate.

4.4.4 Migration Test Method

Ollivier et al. (1997) and Castellote et al. (1999a) determined the binding isotherm after the migration test. Castellote et al. (1999a) determined the amount of bound chlorides from various layers of a concrete sample after it was subjected to a non-steady state migration experiment. In the experiment, X-ray fluorescence was used to obtain the total amount of chlorides, and a leaching method was used to determine the free amount of chlorides.

Ollivier et al. (1997) used steady state migration tests to obtain chloride binding isotherm. After a steady state migration test, the chloride distribution was almost constant throughout the specimen. One point on the binding isotherm can be measured from one migration test. By conducting a series of migration test with different upstream chloride concentrations, a full binding isotherm could be obtained.

Electrical field shortens the time to reach the equilibrium. However, whether the application of electrical field has an influence on the chloride binding property of cement is still not fully understood.

As described, several different methods have been used to determine the free and bound chloride concentration. Besides the four methods described above, other methods to obtain or predict the chloride binding isotherm of cement-based materials are also applied (Baroghel-Bouny et al. 2012; Ramírez-Ortíz et al. 2018). Each of them uses different procedures and has some drawbacks. This in turn makes the comparison between different experimental results difficult. On the other hand, chloride binding can be a very important factor in predicting the service life of concrete. Thus, it is important to standardize one procedure so the results can be comparable and can be incorporated into the service life prediction models.

4.5 DETERMINATION OF PHYSICALLY ABSORBED CHLORIDE DISTRIBUTION IN EDL

Knowledge of water distribution of hardened cement pastes is necessary to explain the chemical and physical properties of cement-based materials. The penetration of aggressive substances such as chloride ions and carbon dioxide is of great importance on degradation of concrete construction, while water plays a remarkable role as a medium for the transportation of these species within materials. According to Powers' model presented by Powers and Brownyard (1946) from a comprehensive study of water vapor sorption isotherms and chemically bound water in hardening cement pastes, water held in cement-based materials can be classified into three categories: capillary water (free water), gel water (physically bound water), and chemically nonevaporable water (bound water).

Chemically bound water has been widely studied with the aid of measurement techniques, such as thermogravimetric analysis (TGA) (Pane and

Hansen 2005) and Quasielastic neutron scattering (QNS) measurements (Berliner et al. 1998). Due to the non-mobility of chemically bound water, the influences of this type of water on performance of cement-based materials are generally insignificant. However, the free and adsorbed water plays an important role in the transportation of aggressive substances and interfacial properties of cement-based materials. Based on the Stern model about EDL formed at the solid–liquid interface, the adsorbed water in EDL has different properties than that in bulk pore solution. From the electrochemical point of view, the water molecules were considered being accumulated together with the charged ions within sub-nanometer distance from charged electrodes (Feng et al. 2014). The accumulation of water molecules is governed by the association of water molecules with their surrounding ions, which drives water molecules into position where the ions are highly charged. Bager et al. (1986a,b) studied the ice formation in hardened cement pastes with low-temperature Calvet microcalorimeter. They found that water contained in finer pores, physically adsorbed water in EDL formed on solid surface, or in a "interlayer" position was nonfrozen even under the condition of $-55°C$, and the freezable (free) water within saturated cement paste can be removed at a relative vapour pressure lower than 60% of atmospheric pressure.

The adsorption of water molecules and charged ions in EDL formed at the solid–liquid interface has been widely studied for illustration of the ion migration and interfacial properties of cement-based materials. In 1991, Hawes and Feldman (1992) studied the adsorption of organic-phase change materials in concrete buildings for heat storage, and demonstrated that factors including concrete structure, temperature, viscosity of liquid, immersion duration, and adsorption area all affected the amount of phase change materials which can be adsorbed. Friedmann et al. (Friedmann et al. 2008; Nguyen and Amiri 2014) studied the effects of EDL on chloride and multispecies ion transport in cement-based materials by a proposed physical EDL model. In these papers, the formation of EDL, overlapping of EDL, distribution of ions in EDL, and how these affect transport of chloride ions or other ions were theoretically studied and discussed. Especially for chloride penetration, the existence of EDL on the surface of solid wall was considered to attract chloride ions onto the surface of cement hydration products and then released into the expressed pore solution and increased the chloride concentration. It has been investigated (Shi 1992) that the freezing of water in a hardened cement paste is not a continuous process and the capillary water may become ice in a specific temperature range, while for water adsorbed on the surface of solid or fixed in small pore, they mostly belong to non-frozen water. However, due to the absence of measuring techniques, the formation and properties of EDL within cement-based materials failed to be verified by experiment.

Nuclear magnetic resonance (NMR) ¹H relaxometry is a non-destructive technique to investigate the content and distribution of proton (H) within samples. ¹H NMR has been shown to be a powerful tool for the characterization of pore size distribution and porosity of cement pastes at the nanoscale owing to the proportional correlation between the ^{1}H relaxation rate of water in pore spaces and the pore surface to volume ratio (Gajewicz et al. 2016). In geotechnical engineering, ^{1}H NMR has been applied in detecting pore distribution and adsorbed water content. Tian et al. (Tian and Wei 2014; Tian et al. 2014) developed a method to distinguish the free and adsorbed water of clay based on the different freezing points and resistances to suction between adsorbed and free water by ^{1}H NMR. Compared to clay or other soil materials, cement-based materials have a denser structure and lower total porosity, which makes it more difficult to separate the free and adsorbed water.

In this part, a new approach to determine the content of adsorbed water within pores of cement paste by ^{1}H NMR relaxometry test is introduced (Hu 2017). The boundary value of T2 relaxation time for capillary (free) and adsorbed water was determined based on the variation of freezing point. Even further studies are needed to verify the availability of this method in the characterization of adsorbed water and ions in cement-based materials; the results obtained in this study may provide some information for further understanding and studying in interfacial properties, EDL formation, and chloride absorption of cement-based materials.

^{1}H NMR relaxation tests were conducted for cement pastes after soaking in NaCl solutions with 0.1, 0.3, 0.5, 0.7, and 1.0 mol/L for 91 days. Small fragments with around 0.5 cm diameter were used for NMR test. Niumag MicroMR12-025 was used for NMR relaxometry measurement; this NMR instrument can measure the T2 relaxation time distribution of samples at temperature ranging from –30 to 30°C. The NMR test was done at ambient temperature first, then put into a cold bath from 5°C decreased to –30°C at a step of 5.0°C. For every point, the sample was kept in the cold bath for 24 hours and then shifted to the sample tube for NMR test. During the test, cryogenic nitrogen gas equal to the temperature of the cold bath was flowed around the sample tube to avoid the variation of temperature inside the samples. The testing sample was put back to the cold bath after the NMR test and applied for the temperature next step. The resonance frequency was 11.845 MHz and Carr–Purcell–Meiboom–Gill (CPMG) pulse sequence was employed to measure NMR transverse relaxation time. The repetition time of sampling was 100 ms, number of echo 1000 and half echo time 120 ms. The free induction decay (FID) curve of samples were then inversed (inversion software provided by Niumag corporation was applied) into the T2 relaxation time distribution at different temperature.

Figure 4.5 shows the evolution of T2 relaxation distribution within cement paste samples with the decrease of temperature from 5 to –30 °C; experimental results of cement pastes immersed in 0.1, 0.5, and 1.0 mol/L NaCl solutions are presented. It can be seen from the figures that the water in large pores gradually turned into ice. At the range of 1,000 to 10,000 ms, T2 signal was only detected for test under the temperature of –5, 0, and 5°C. Compared to the peak at the range of 0.01 to 1.0 ms, the T2 signal on

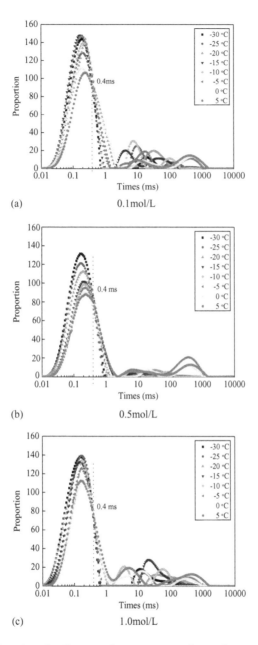

Figure 4.5 T2 relaxation distribution of cement paste during freezing process. (From Hu 2017.)

the right hand side of the T2 relaxation distribution plot is relatively small, especially for cement paste under lower temperature. When the temperature decreased to –30 °C, nearly 95% of signals were located at the range of 0.01 to 1 ms. However, for the T2 signal peaks with T2 relaxation time in this section, it can be seen that the signal amplitude with relaxation time lower than 0.4 ms increased with the decrease of temperature, while the signal decreased when relaxation time higher than 0.4 ms.

The value of T2 in NMR measurement results can represent the location of water within samples. High T2 value relates to water in large pores, while the water in small pores and the adsorbed water can result into low T2 value. In this study, we can easily regard the peak within 0.01 to 1.0 ms as the water in small pores and adsorbed water. Generally, the liquid in a pore has a frozen temperature related to the diameter of pore. Jehng et al. (1996) studied the microstructural evolution of cement paste by NMR and freezing; it was found that the capillary pore water can be frozen at temperature lower than –30°C. However, water remained as liquid in gel pores over a temperature range reaching to –120°C. The confinement of solid phase or solid surface on ions and liquid which may change the existence form of solution within pore structure was considered as the differences in freezing temperature.

It can be seen in this study that the water at the range of 0.01 to 1 ms remained being liquid; the results in this section can be regarded as the water distribution before freezing. The effects of surface potential on liquids and ions in EDL result in different properties of adsorbed water in EDL and free water in bulk pore. Due to the surface potential of solid phase, solution in EDL can be only partially moved parallel to solid phase, but not in the direction perpendicular to surface. In this study, the total content of water was unchanged, and we assumed that the water will not be transferred between small and large pores during the freezing of water in large pores. It has been investigated (Yong 1962) that the content of adsorbed water can be changed with different temperature. In this study, variation of adsorbed and free water should be opposite because the total water content was unchanged. Therefore, it can be seen from the experimental results that the content of adsorbed water increases while free water decreases with the decrease of temperature. According to the T2 relaxation distribution plots of cement pastes in different concentrations of NaCl solution shown in Figure 4.5, it can be seen the boundary points of T2 relaxation time for adsorbed and free water of these three samples are basically same. On the two sides of 0.4 ms, the T2 relaxation distribution plots show different trends with different temperature, which means the signal on these two sides presents two different types of pore water. Therefore, we determined T2 = 0.4 ms as the boundary point for adsorbed and free water.

Based on NMR technique, adsorbed and free water has been distinguished by detecting the T2 relaxation distribution under low temperature or different suction. For some materials with higher porosity and water content, such as clay, it has been found that only the peak with T2 relaxation time lower than a boundary value can be detected under low temperature and high suction (Tian et al. 2014). This value was considered to distinguish the adsorbed and free water within pores. However, this study failed to turn all of the free water into ice because −30°C is the lowest temperature in the NMR probe that could be reached. It's difficult to extract water within cement paste by external applied suction. Further studies are needed to verify the boundary T2 relaxation time obtained in this study.

After the boundary point in T2 distribution plots for adsorbed and free water is determined, the percentage of adsorbed water within cement paste under ambient temperature (20°C) can be calculated. The peak area of T2 distribution represents the content of water, then the area on plot on left hand side of 0.4 ms was determined and divided by the total area of distribution plot. Figure 4.6 shows the percentage of adsorbed water to total water within cement pastes as a function of chloride concentration in soaking solution. It can be seen from the figure that the percentage of adsorbed water in cement pastes gradually decreases with the increase of concentration in pore solution (soaking solution). For cement paste immersed in 0.1 mol/L NaCl solution, almost 20% of total water was adsorbed onto the surface of solid phase, while it decreased 5% as concentration increased to 1.0 mol/L.

Our previous studies have examined the influences of chloride concentration or ion concentration in soaking solution on ion distribution and content within EDL. The increased ion concentration in pore solution compressed the range of EDL and decreased the thickness of EDL. According to the Debye formula, the thickness of EDL is significantly decreased with the increase of pore solution concentration. Within the inner pores of cement pastes, the water can be grouped into adsorbed water confined in EDL and free water in bulk pore. The increase of EDL thickness can raise the content of adsorbed water, and a similar trend between EDL thickness

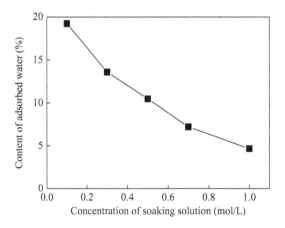

Figure 4.6 Content of adsorbed water in cement paste. (From Hu 2017.)

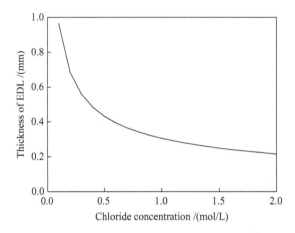

Figure 4.7 Calculated thickness of EDL according to Debye formula. (From He 2010.)

and adsorbed water content with concentration of soaking solution can be found in Figure 4.6 and calculated results according to Debye formula (Figure 4.7) (He 2010). Apart from the concentration of soaking solution, the pore structure of cement paste will also affect the percentage of adsorbed water to total water content. The reaction of chloride ions and solid phase can decrease the porosity of cement-based materials and increase the density. With a fixed EDL thickness, the small diameter pore reveals higher percentage of value of EDL area to that of total pore and adsorbed water to total water in pore. Therefore, comparing the plots in Figures 4.6 and 4.7, the decrease of EDL thickness with concentration of soaking solution is more obvious than the evolution of adsorbed water content.

4.6 EFFECT OF CHLORIDE BINDING ON MICROSTRUCTURE

4.6.1 Effects of Chloride Binding on Hydration Products

Koleva et al. (2007) investigated the variations of chemical composition of hydration products in cement-based materials with internal chloride salts. As shown in Figure 4.8, the C/S ratio of C-S-H gel in samples with chloride ions was between 2.19 to 2.95, while C/S ratio around 1.8 was found in samples without chloride ions. The surface potential of C-S-H gel is mainly determined by the chemical composition, especially C/S ratio of C-S-H gel. For high C/S ratio C-S-H gel, the surface is positively charged and anion ions such as Cl⁻ and OH⁻ in pore solution can be absorbed (Monteiro et al. 1997). Oppositely, as the C/S ratio of C-S-H gel decreases to lower than 1.2-1.3,

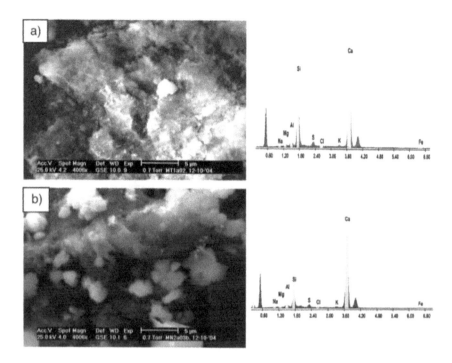

Figure 4.8 SEM images and the corresponding EDX spectrums of plain mortars: (a) without admixed chlorides and (b) with admixed chlorides. (From Koleva et al. 2007.)

its surface will be negatively charged and cation ions such as Na^+ and K^+ in pore solution can be absorbed and remain Cl^- in pore solution.

Besides C-S-H gel, chloride binding can also have some effects on other hydration products. Ekolu et al. (2006) reported that at a specific chloride concentration, AFm phase in cement-based materials would be destroyed while most of the AFt phase remained unchanged. As the chloride concentration continually increased, the AFm and AFt phases were both decomposed and transformed into Friedel's salt and gypsum. Balonis et al. (2010) studied the interaction between chloride ions and AFm phases within hydrated cement, and a comprehensive picture of phase transfer with the increase of chloride concentration was provided, as shown in Figure 4.9. When chloride ions are introduced under service conditions, chloride ions can readily displace the OH, SO_4, and CO_3 ions in AFm phases and transfer into Kuzel's salt or Friedel's salt. Meanwhile, ettringite with lower density is formed and results in the increase of solid phase volume (Jensen and Pratt 1989; Shi et al. 2016).

4.6.2 Effects of Chloride Binding on Pore Structure

The pore structure of cement-based materials will also be affected by intruded or penetrated chloride ions (Wang et al. 2013). Midgley and

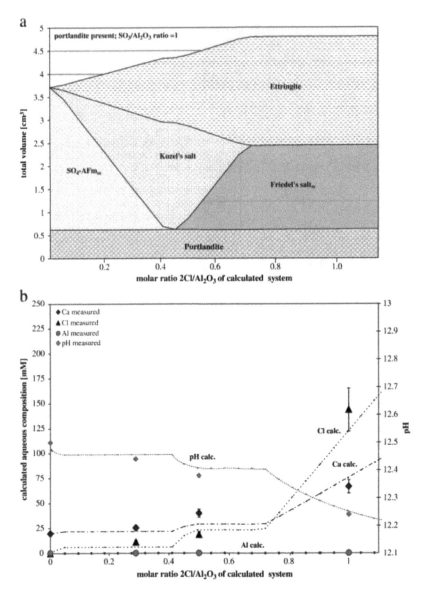

Figure 4.9 (a) Calculated total volume, (b) calculated and experimental aqueous composition of hydrated model mixture. (Balonis et al. 2010.)

Illston (1986) studied the pore size distribution of cement paste after the penetration of chloride ions and found that the introduced chloride ions resulted in the increase of fine pores and the decrease of large pores. Jensen et al. (Jensen and Pratt 1989; Suryavanshi and Swamy 1998) also found that the Friedel's salt formed within the reaction between cement hydrates and chloride ions precipitated in large pores and decreased the porosity and permeability of concrete, while for samples with 10% silica fume, the effects of chloride binding on pore structure were not that obvious. For samples with lower w/b ratio, the improvement of chloride binding on pore size distribution will be more significant (Zhang and Gjørv 1991).

Diaz et al. (2008) studied the microstructural changes of cement mortar exposed in vacuum saturation of sodium chloride solution by AC impedance spectroscopy. The results showed that in the process of chloride ion introduction, the microstructure of cement mortar was changed due to formation of Friedel's salt. Meanwhile, the porosity and tortuosity of samples were increased, and a new pore structure system formed within cement mortar. The increase of porosity was ascribed to the calcium ion leaching of cement-based materials at lower external chloride ion concentration. Also, by using AC impedance spectroscopy technology, Sánchez et al. (Jain and Neithalath 2010; Sánchez et al. 2008) studied the variations of pore structure of concrete during RCM test. It was found that the pore size was gradually decreased during ion migration, which was in agreement with MIP test results. As shown in Figure 4.10, the pore network of concrete experiments a narrowing-down process; a tendency to decrease the diameters of the pores was obtained. The formation of a new solid compound was considered as the reason to this variation.

In conclusion, the penetration of chloride ions into cement-based materials improves the pore structure mainly in decrease of pore size and increase of tortuosity. The formation of a new solid phase Friedel's salt and

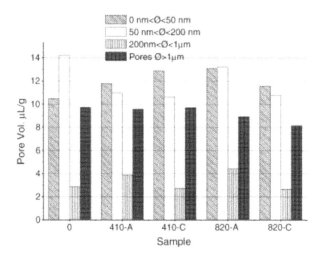

Figure 4.10 Measured pore volumes at four pore diameter ranges, see text for details. The data correspond to a reference concrete sample (0), not subjected to migration, and to samples obtained after 410 and 820 h of migration for both anodic and cathodic sides. (From Sánchez et al. 2008.)

adsorption of chloride ions on pore wall or surface of C-S-H gel are the main reasons.

Compared to chloride diffusion test, the applications of high electrical voltage in rapid chloride penetration (RCP) and RCM tests speed up the ingress of chloride ions within cement-based materials. Due to the different testing periods and temperature variation compared to natural diffusion test, it is hard to determine the microstructural variation during RCM test. In some studies (Balonis et al. 2010; Page et al. 1981), it was shown that the electrical resistance and mass of cement mortar gradually increased during the RCM test. According to our recent studies, as shown in Figure 4.11, the pore structure of cement paste was improved with the penetration of chloride into sample, especially after rapid migration test. The Ca/Si of C-S-H gel was increased compared to samples after natural diffusion test and samples exposed to non-chloride environment. The applied voltage modified the size and shape of Friedel's salt formed within cement pastes.

Studies on effects of chloride binding on microstructure of cement-based materials during chloride migration test are also limited. The time interval between chemical reaction in pore solution and in catholyte and anolyte solution is very short, which increases the difficulty of analysis on microstructure during this process. Jain and Neithalath (2011) quantitatively studied the microstructural changes of concrete during non-steady state chloride migration test by AC impedance spectroscopy. A pore structure parameter $\beta\Phi$ was introduced to represent the microstructure of concrete, where Φ is the porosity of samples and β represents the connectivity of pores. The results showed that the value of $\beta\Phi$ of concrete on average decreased 10% during the non-steady state chloride migration test. The results indicated that during the migration test, the porosity and content of connected pores of concrete decreased, while the pore system became more complicated. More studies are needed in this field, especially when more than one kind of mineral admixture is applied. Meanwhile, it can be also found that AC impedance spectroscopy is an effective method to investigate the chloride transportation and microstructural variations of cement-based materials. Wu and Yan (2012) reviewed the application of AC impedance spectroscopy in studying chloride diffusion process in cement-based materials, which laid a foundation on wide application of this technique in cement-based materials.

4.7 SUMMARY

Chloride binding is a very complicated process and is affected by many factors, such as chloride concentration, cement composition, hydroxyl concentration, cation of chloride salt, temperature, supplementary cementing materials, carbonation, sulfate ions, and electrical field. In many cases, it is assumed that concrete is saturated and carbonation does not happen.

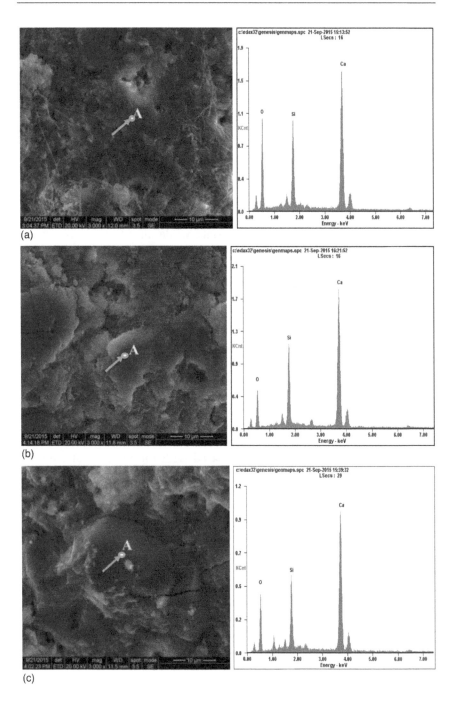

Figure 4.11 SEM and EDX results of samples (a) in non-chloride environment, (b) after natural diffusion and (c) RCM tests.

In fact, in some cases, such as splash zone and concrete subjected to de-icing salt, carbonation may occur progressively at the same time. However, there are few publications which deal with the effect of carbonation on the binding. It is necessary to establish the relationship between carbonation and binding.

An electrical field is widely used to accelerate chloride transport. However, the effect of an electrical field on the binding behavior of cement-based materials is generally neglected. The binding in the presence of an electrical field may differ from the binding of natural diffusion in the binding capacity and binding rate. To use the results from rapid migration tests to predict the chloride transport in concrete, the difference in chloride binding between diffusion and migration should be understood.

The replacement of cement with GGBFS or fly ash increases the chloride binding capacity of cement-based materials, while silica fume decreases the chloride binding. C3A plays the most important role in chloride binding. C3S, C2S, and C4AF contribute to chloride binding as well. Predicting chloride binding from the composition of cement is always desired. Published data already show that this is quite promising. Although a lot of data have been published on the effect of the composition of cement and supplementary cementing materials on the chloride binding, they are not good enough to develop a model which can predict the chloride binding from the composition of cement. More comprehensive data on the effect of the composition of cement and supplementary cementing materials on the binding are needed.

A new approach to determine the adsorbed water content in EDL is introduced. The adsorbed and free water in bulk pore can be distinguished by NMR based on different variations of two during freezing process. The percentage of adsorbed water is decreased with the increase of chloride concentration in soaking solution due to the decrease of EDL thickness.

No single binding isotherm can accurately express the relationship between free and bound chloride within the complete concentration range. Linear binding seems to oversimplify the relationship between free and bound chloride. The assumption of monolayer adsorption in the Langmuir isotherm seems far from reality at high concentrations, although it fits well with experimental results at low concentrations. Due to the numerical complexity, BET isotherm is seldom applied in the service life prediction models. The Freundlich isotherm corresponds with the experimental results very well at the concentrations which cover the range of free chloride concentrations in seawater. When no binding or linear binding is considered in the prediction models, it underestimates the service life.

However, in some long-term field cases, a linear relationship between free and bound chloride was found. This may be ascribed to the leakage of hydroxyl. So, in order to accurately describe the relationship between free and bound chloride in field cases, many other factors, such as the leakage of hydroxyl, temperature, etc., should be considered.

Modeling binding phenomena by chemical equilibrium at each point is advanced over binding isotherm techniques. However, the physical binding should not be neglected. More research is needed to develop models incorporating physical binding.

REFERENCES

Ann KY, Hong SI. Modeling chloride transport in concrete at pore and chloride binding. *ACI Materials Journal*. 2018, 115(4):595-604.

Arya C, Buenfeld N, Newman J. Factors influencing chloride-binding in concrete. *Cement and Concrete Research*. 1990, 20:291–300.

Arya C, Xu Y. Effect of cement type on chloride binding and corrosion of steel in concrete. *Cement and Concrete Research*. 1995, 25:893–902.

Azad VJ, Isgor OB. A thermodynamic perspective on admixed chloride limits of concrete produced with SCMs. Special Publication. 2016, 308:1–18.

Bager DH, Sellevold EJ. Ice formation in hardened cement paste. Part I—Room temperature cured pastes with variable moisture contents. *Cement and Concrete Research*. 1986a, 16:709–720.

Bager DH, Sellevold EJ. Ice formation in hardened cement paste. Part II—Drying and resaturation on room temperature cured pastes. *Cement and Concrete Research*. 1986b, 16:835–844.

Balonis M, Lothenbach B, Le Saout G, Glasser FP. Impact of chloride on the mineralogy of hydrated Portland cement systems. *Cement and Concrete Research*. 2010, 40:1009–1022.

Barbarulo R, Marchand J, Snyder KA, Prené S. Dimensional analysis of ionic transport problems in hydrated cement systems: Part 1. Theoretical considerations. *Cement and Concrete Research*. 2000, 30:1955–1960.

Baroghel-Bouny V, Wang X, Thiery M, Saillio M, Barberon F. Prediction of chloride binding isotherms of cementitious materials by analytical model or numerical inverse analysis. *Cement and Concrete Research*. 2012, 42:1207–1224.

Beaudoin JJ, Ramachandran VS, Feldman RF. Interaction of chloride and C-S-H. *Cement and Concrete Research*. 1990, 20:875–883.

Berliner R, Popovici M, Herwig K, et al. Quasielastic neutron scattering study of the effect of water-to-cement ratio on the hydration kinetics of tricalcium silicate. *Cement and Concrete Research*. 1998, 28:231–243.

Bigas J-P. *La diffusion des ions chlore dans les mortiers*. INSA, Toulouse; 1994.

Blunk G, Gunkel P, Smolczyk HG. On the distribution of chloride between the hardening cement pates and its pore solutions. Proceedings of the 8th International Congress on the Chemistry of Cement, Rio de Janeiro, Brazil, 1986, 4:85–90.

Brown P, Badger S. The distributions of bound sulfates and chlorides in concrete subjected to mixed NaCl, MgSO4, Na2SO4 attack. *Cement and Concrete Research*. 2000, 30:1535–1542.

Byfors K. Chloride binding in cement paste. *Nordic Concrete Research*. 1986, 5:27–38.

Cao Y, Guo L, Chen B. Influence of sulfate on the chloride diffusion mechanism in mortar. *Construction and Building Materials*. 2019, 197:398–405.

Castellote M, Andrade C, Alonso C. Chloride-binding isotherms in concrete submitted to non-steady-state migration experiments. *Cement and Concrete Research*. 1999, 29:1799–1806.

Castellote M, Andrade C, Alonso C. Measurement of the steady and non-steady-state chloride diffusion coefficients in a migration test by means of monitoring the conductivity in the anolyte chamber. Comparison with natural diffusion tests. *Cement and Concrete Research*. 2001, 31:1411–1420.

Chang H. Chloride binding capacity of pastes influenced by carbonation under three conditions. *Cement and Concrete Composites*. 2017, 84:1–9.

Cheewaket T, Jaturapitakkul C, Chalee W. Long term performance of chloride binding capacity in fly ash concrete in a marine environment. *Construction and Building Materials*. 2010, 24:1352–1357.

Díaz B, Freire L, Merino P, Novoa X, Pérez M. Impedance spectroscopy study of saturated mortar samples. *Electrochimica Acta*. 2008, 53:7549–7555.

De Weerdt K, Orsáková D, Geiker MR. The impact of sulphate and magnesium on chloride binding in Portland cement paste. *Cement and Concrete Research*. 2014, 65:30–40.

Delagrave A, Marchand J, Ollivier J-P, Julien S, Hazrati K. Chloride binding capacity of various hydrated cement paste systems. *Advanced Cement based Materials*. 1997, 6:28–35.

Dhir R, El-Mohr M, Dyer T. Developing chloride resisting concrete using PFA. *Cement and Concrete Research*. 1997, 27:1633–1639.

Dhir R, Hubbard F, Unsworth H. XRF thin film copper disc evaporation test for the elemental analysis of concrete test solutions. *Cement and Concrete Research*. 1995, 25:1627–1632.

Dousti A, Shekarchi M, Alizadeh R, Taheri-Motlagh A. Binding of externally supplied chlorides in micro silica concrete under field exposure conditions. *Cement and Concrete Composites*. 2011, 33:1071–1079.

Ekolu S, Thomas M, Hooton R. Pessimum effect of externally applied chlorides on expansion due to delayed ettringite formation: Proposed mechanism. *Cement and Concrete Research*. 2006, 36:688–696.

Feng G, Jiang X, Qiao R, Kornyshev AA. Water in ionic liquids at electrified interfaces: The anatomy of electrosorption. *ACS Nano*. 2014, 8:11685–11694.

Florea M, Brouwers H. Modelling of chloride binding related to hydration products in slag-blended cements. *Construction and Building Materials*. 2014, 64:421–430.

Frías M, Goñi S, García R, et al. Seawater effect on durability of ternary cements. Synergy of chloride and sulphate ions. *Composites Part B: Engineering*. 2013, 46:173–178.

Friedmann H, Amiri O, Aït-Mokhtar A. Physical modeling of the electrical double layer effects on multispecies ions transport in cement-based materials. *Cement and Concrete Research*. 2008, 38:1394–1400.

Gajewicz A, Gartner E, Kang K, McDonald P Yermakou V. A 1 H NMR relaxometry investigation of gel-pore drying shrinkage in cement pastes. *Cement and Concrete Research*. 2016, 86:12–19.

Gardner TJ. *Chloride Transport Through Concrete and Implications for Rapid Chloride Testing*. University of Cape Town; 2006.

Geng J, Easterbrook D, Li L, et al. The stability of bound chlorides in cement paste with sulfate attack. *Cement and Concrete Research*. 2015, 68:211–222.

Glass G, Buenfeld N. The influence of chloride binding on the chloride induced corrosion risk in reinforced concrete. *Corrosion Science*. 2000, 42:329–344.

Glass G, Stevenson G, Buenfeld N. Chloride-binding isotherms from the diffusion cell test. *Cement and Concrete Research*. 1998, 28:939–945.

Glass G, Wang Y, Buenfeld N. An investigation of experimental methods used to determine free and total chloride contents. *Cement and Concrete Research*. 1996, 26:1443–1449.

Hassan Z. *Binding of External Chloride by Cement Pastes*. University of Toronto, Toronto, ON; 2001.

Hawes D, Feldman D. Absorption of phase change materials in concrete. *Solar Energy Materials and Solar Cells*. 1992, 27:91–101.

He F. *Measurement of Chloride Migration in Cement-based Materials Using AgNO3 Colorimetric Method*. Central South University, Changsha; 2010.

He F, Shi C, Hu X, et al. Calculation of chloride ion concentration in expressed pore solution of cement-based materials exposed to a chloride salt solution. *Cement and Concrete Research*. 2016, 89:168–176.

Hu X. *Mechanism of Chloride Concentrate and its Effects on Microstructure and Electrochemical Properties of Cement-Based Materials*. Ghent University; 2017.

Hussain SE, Al-Gahtani AS. Pore solution composition and reinforcement corrosion characteristics of microsilica blended cement concrete. *Cement and Concrete Research*. 1991, 21:1035–1048.

Ipavec A, Vuk T, Gabrovšek R, Kaučič V. Chloride binding into hydrated blended cements: The influence of limestone and alkalinity. *Cement and Concrete Research*. 2013, 48:74–85.

Jain J, Neithalath N. Electrical impedance analysis based quantification of microstructural changes in concretes due to non-steady state chloride migration. *Materials Chemistry and Physics*. 2011, 129:569–579.

Jain J, Neithalath N. Chloride transport in fly ash and glass powder modified concretes–influence of test methods on microstructure. *Cement and Concrete Composites*. 2010, 32:148–156.

Jehng J-Y, Sprague D, Halperin W. Pore structure of hydrating cement paste by magnetic resonance relaxation analysis and freezing. *Magnetic Resonance Imaging*. 1996, 14:785–791.

Jensen H-U, Pratt P. The binding of chloride ions by pozzolanic product in fly ash cement blends. *Advances in Cement Research*. 1989, 2:121–129.

Jensen OM, Korzen M, Jakobsen H, Skibsted J. Influence of cement constitution and temperature on chloride binding in cement paste. *Advances in Cement Research*. 2000, 12:57–64.

Jung MS, Kim KB, Lee SA, et al. Risk of chloride-induced corrosion of steel in SF concrete exposed to a chloride-bearing environment. *Construction and Building Materials*. 2018, 166:413–422.

Kayali O, Khan M, Ahmed MS. The role of hydrotalcite in chloride binding and corrosion protection in concretes with ground granulated blast furnace slag. *Cement and Concrete Composites*. 2012, 34:936–945.

Khan MSH, Kayali O. Chloride binding ability and the onset corrosion threat on alkali-activated GGBFS and binary blend pastes. *European Journal of Environmental and Civil Engineering*. 2018, 22(8):1023–1039.

Khan MSH, Kayali O, Troitzsch U. Chloride binding capacity of hydrotalcite and the competition with carbonates in ground granulated blast furnace slag concrete. *Materials and Structures*. 2016, 49:4609–4619.

Kim MJ, Kim KB, Ann KY. The influence of C3A content in cement on the chloride transport. *Advances in Materials Science and Engineering*. 2016, 2016.

Kopecskó K, Balázs GL. Concrete with improved chloride binding and chloride resistivity by blended cements. *Advances in Materials Science and Engineering*. 2017, 2017.

Koleva D, Hu J, Fraaij A, Van Breugel K, De Wit J. Microstructural analysis of plain and reinforced mortars under chloride-induced deterioration. *Cement and Concrete Research*. 2007, 37:604–617.

Krishnakumark BP. Evaluation of chloride penetration in OPC concrete by silver nitrate solution spray method. *International Journal of ChemTech Research*. 2014, 6:2676–2682.

Lambert P, Page C, Short N. Pore solution chemistry of the hydrated system tricalcium silicate/sodium chloride/water. *Cement and Concrete Research*. 1985, 15:675–680.

Larsen C. Chloride binding in concrete, Effect of surrounding environment and concrete composition. These de doctorat, The Norwegian University of Science and Technology Trondheim Norway, 1998.

Larsson J. The enrichment of chlorides in expressed concrete pore solution submerged in saline solution. Proceedings of the Nordic seminar on field studies of chloride initiated reinforcement corrosion in concrete, Lund University of Technology, Report TVBM-3064, 1995:171–176.

Liu J, Qiu Q, Chen X, et al. Degradation of fly ash concrete under the coupled effect of carbonation and chloride aerosol ingress. *Corrosion Science*. 2016a, 112:364–372.

Liu J, Qiu Q, Chen X, et al. Understanding the interacted mechanism between carbonation and chloride aerosol attack in ordinary Portland cement concrete. *Cement and Concrete Research*. 2017, 95:217–225.

Liu W, Cui H, Dong Z, et al. Carbonation of concrete made with dredged marine sand and its effect on chloride binding. *Construction and Building Materials*. 2016b, 120:1–9.

Liu Y, Shi X. Electrochemical chloride extraction and electrochemical injection of corrosion inhibitor in concrete: State of the knowledge. *Corrosion Reviews*. 2009, 27:53–82.

Ma B, Mu S, De Schutter G. Non-steady state chloride migration and binding in cracked self-compacting concrete. *Journal of Wuhan University of Technology-Mater. Sci. Ed.* 2013, 28:921–926.

Machner A, Zajac M, Haha MB, et al. Chloride-binding capacity of hydrotalcite in cement pastes containing dolomite and metakaolin. *Cement and Concrete Research*. 2018, 107:163–181.

Maes M, Gruyaert E, De Belie N. Resistance of concrete with blast-furnace slag against chlorides, investigated by comparing chloride profiles after migration and diffusion. *Materials and Structures*. 2013, 46:89–103.

Mehta P. Effect of cement composition on corrosion of reinforcing steel in concrete. *Chloride Corrosion of Steel in Concrete*. ASTM International; 1977.

Midgley H, Illston J. Effect of chloride penetration on the properties of hardened cement pastes. 7th International Congress on the Chemistry of Cement, 1986, 101–103.

Mohammed T, Hamada H. Relationship between free chloride and total chloride contents in concrete. *Cement and Concrete Research*. 2003, 33:1487–1490.

Monteiro P, Wang K, Sposito G, Dos Santos M, de Andrade WP. Influence of mineral admixtures on the alkali-aggregate reaction. *Cement and Concrete Research*. 1997, 27:1899–1909.

Nagataki S, Otsuki N, Wee T-H, Nakashita K. Condensation of chloride ion in hardened cement matrix materials and on embedded steel bars. *Materials Journal*. 1993, 90:323–332.

Nguyen P, Amiri O. Study of electrical double layer effect on chloride transport in unsaturated concrete. *Construction and Building Materials*. 2014, 50:492–498.

Nilsson L, Poulsen E, Sandberg P, Sørensen H, Klinghoffer O. HETEK. Chloride penetration into concrete. *State of the Art, Transport Processes, Corrosion Initiation, Test Methods and Prediction Models*. Technical report in The Road Directorate, Denmark, 1996. ISSN/ISBN:0909–4288.

Oh BH, Jang SY. Effects of material and environmental parameters on chloride penetration profiles in concrete structures. *Cement and Concrete Research*. 2007, 37:47–53.

Olivier T. Prediction of chloride penetration into saturated concrete—multi-species approach: PhD thesis, Department of Building Materials, Chalmers University of Technology, Goteborg, Sweden, 2000.

Ollivier J, Arsenault J, Truc O, Marchand J. Determination of chloride binding isotherms from migration tests. Mario Collepardi Symposium on Advances in Concrete Science and Technology, Rome, 1997, 198–217.

Page C, Lambert P, Vassie P. Investigations of reinforcement corrosion. 1. The pore electrolyte phase in chloride-contaminated concrete. *Materials and Structures*. 1991, 24:243–252.

Page C, Short N, El Tarras A. Diffusion of chloride ions in hardened cement pastes. *Cement and Concrete Research*. 1981, 11:395–406.

Page C, Vennesland Ø. Pore solution composition and chloride binding capacity of silica-fume cement pastes. *Matériaux et Construction*. 1983, 16:19–25.

Pane I, Hansen W. Investigation of blended cement hydration by isothermal calorimetry and thermal analysis. *Cement and Concrete Research*. 2005, 35:1155–1164.

Papadakis VG. Effect of supplementary cementing materials on concrete resistance against carbonation and chloride ingress. *Cement and Concrete Research*. 2000, 30:291–299.

Potgieter JH, Delport D, Verryn S, Potgieter-Vermaak S. Chloride-binding effect of blast furnace slag in cement pastes containing added chlorides. *South African Journal of Chemistry*. 2011, 64:108–114.

Powers TC, Brownyard TL. Studies of the physical properties of hardened Portland cement paste. *Journal Proceedings*. 1946:101–132.

Qiao C, Ni W, Wang Q, et al. Chloride diffusion and wicking in concrete exposed to NaCl and MgCl2 solutions. *Journal of Materials in Civil Engineering*. 2018, 30(3):04018015.

Ramachandran VS. Possible states of chloride in the hydration of tricalcium silicate in the presence of calcium chloride. *Matériaux et Construction*. 1971, 4:3–12.

Ramachandran VS, Seeley R, Polomark G. Free and combined chloride in hydrating cement and cement components. *Matériaux et Construction*. 1984, 17:285–289.

Ramírez-Ortíz AE, Castellanos F, Cano-Barrita PFJ. Ultrasonic detection of chloride ions and chloride binding in portland cement pastes. *International Journal of Concrete Structures and Materials*. 2018, 12(1):20. Rasheeduzzafar DFH. Influence of cement composition on concrete durability. *ACI Material Journal*. 1992, 89(6):574–85.

Roberts M. Effect of calcium chloride on the durability of pre-tensioned wire in prestressed concrete. *Magazine of Concrete Research*. 1962, 14:143–154.

Sánchez I, Nóvoa X, De Vera G, Climent M. Microstructural modifications in Portland cement concrete due to forced ionic migration tests. Study by impedance spectroscopy. *Cement and Concrete Research*. 2008, 38:1015–1025.

Saillio M, Baroghel-Bouny V, Barberon F. Chloride binding in sound and carbonated cementitious materials with various types of binder. *Construction and Building Materials*. 2014, 68:82–91.

Saillio M, Bouny VB, Pradelle S. Physical and chemical chloride binding in cementitious materials with various types of binder. 14th International Congress on the Chemistry of Cement, 2015, 12.

Samson E, Marchand J, Snyder KA. Calculation of ionic diffusion coefficients on the basis of migration test results. *Materials and Structures*. 2003, 36:156–165.

Sandberg P. Studies of chloride binding in concrete exposed in a marine environment. *Cement and Concrete Research*. 1999, 29:473–477.

Sandberg P, Larsson J. Chloride binding in cement pastes in equilibrium with synthetic pore solutions. Chloride Penetration into Concrete Structures. Nordic Miniseminar; 1993, 98–107.

Sergi G, Yu S, Page C. Diffusion of chloride and hydroxyl ions in cementitious materials exposed to a saline. *Magazine of Concrete Research*. 1992, 44:63–69.

Shi C. *Activation of Natural Pozzolans, Fly Ashes and Blast Furnace Slag*. Civil Engineering, University of Calgary; 1992.

Shi C, Hu X, Wang X, Wu Z, Schutter Gd. Effects of chloride ion binding on microstructure of cement pastes. *Journal of Materials in Civil Engineering*. 2016, 29:04016183.

Shuguang H, Jian G, Qingjun D. The characters of cement hardened paste with mineral admixtures binding chloride ion under the condition of stray current interfering. *Journal of Huazhong* University of Science and Technology *(Nature Science Edition)*. 2008, 3:011.

Song H-W, Lee C-H, Ann KY. Factors influencing chloride transport in concrete structures exposed to marine environments. *Cement and Concrete Composites*. 2008a, 30:113–121.

Song H, Lee C, Jung M, Ann K. Development of chloride binding capacity in cement pastes and influence of the pH of hydration products. *Canadian Journal of Civil Engineering*. 2008b, 35:1427–1434.

Sotiriadis K, Rakanta E, Mitzithra ME, et al. Influence of sulfates on chloride diffusion and chloride-induced reinforcement corrosion in limestone cement materials at low temperature. *Journal of Materials in Civil Engineering*. 2017, 29(8):04017060.

Spiesz P, Brouwers H. Influence of the applied voltage on the Rapid Chloride Migration (RCM) test. *Cement and Concrete Research*. 2012, 42:1072–1082.

Spiesz P, Brouwers H. The apparent and effective chloride migration coefficients obtained in migration tests. *Cement and concrete Research*. 2013, 48:116–127.

Stanish KD. *The Migration of Chloride Ions in Concrete*. Canada: University of Toronto; 2002.

Sun G-w, Guan X-m, Sun W, Zhang Y-s. Research on the Binding Capacity and Mechanism of Chloride Ion Based on Cement-GGBS System. *Journal of Wuhan University of Technology*. 2010, 7:010.

Suryavanshi A, Scantlebury J, Lyon S. Mechanism of Friedel's salt formation in cements rich in tri-calcium aluminate. *Cement and Concrete Research*. 1996, 26:717–727.

Suryavanshi A, Swamy R. Influence of penetrating chlorides on the pore structure of structural concrete. *Cement, Concrete and Aggregates*. 1998, 20:169–179.

Suryavanshi A, Swamy RN. Stability of Friedel's salt in carbonated concrete structural elements. *Cement and Concrete Research*. 1996, 26:729–741.

Tang L. *Chloride Transport in Concrete-Measurement and Prediction*. Chalmers University of Technology; 1996.

Tang L, Nilsson L-O. Chloride binding capacity and binding isotherms of OPC pastes and mortars. *Cement and Concrete Research*. 1993, 23:247–253.

Thomas M, Hooton R, Scott A, Zibara H. The effect of supplementary cementitious materials on chloride binding in hardened cement paste. *Cement and Concrete Research*. 2012, 42:1–7.

Tian H, Wei C. A NMR-based testing and analysis of adsorbed water content. *Scientia Sinica Technologica*. 2014, 44:295–305.

Tian H, Wei C, Wei H, Zhou J. Freezing and thawing characteristics of frozen soils: Bound water content and hysteresis phenomenon. *Cold Regions Science and Technology*. 2014, 103:74–81.

Tran VQ, Soive A, Bonnet S, et al. A numerical model including thermodynamic equilibrium, kinetic control and surface complexation in order to explain cation type effect on chloride binding capability of concrete. *Construction and Building Materials*. 2018, 191:608–618.

Tritthart J. Chloride binding in cement II. The influence of the hydroxide concentration in the pore solution of hardened cement paste on chloride binding. *Cement and Concrete Research*. 1989, 19:683–691.

Tuutti K. Analysis of pore solution squeezed out of cement paste and mortar. *Nordic Concrete Research*. 1982.

Voinitchi Da, Julien S, Lorente S. The relation between electrokinetics and chloride transport through cement-based materials. *Cement and Concrete Composites*. 2008, 30:157–166.

Vu QH, Pham G, Chonier A, et al. Impact of C3A content on the chloride diffusivity of concrete. *Construction Materials and Systems*. 2017:377.

Wang X, Shi C, He F, et al. Chloride binding and its effects on microstructure of cement-based materials. *Journal of The Chinese Ceramic Society*. 2013, 41:187–198.

Weiss WJ, Isgor OB, Coyle AT, et al. Prediction of chloride ingress in saturated concrete using formation factor and chloride binding isotherm. *Advances in Civil Engineering Materials*. 2018, 7(1):206–220.

Wiens U. Chloride binding of cement paste containing fly ash. Proceedings of the 10th International Congress on the Chemistry of Cement, 1997.

Wowra O, Setzer M, Setzer M, Auberg R. Sorption of chlorides on hydrated cements and C3S pastes. *Frost Resistance of Concrete.* 1997:147–153.

Wu L, Yan P. Review on AC impedance techniques for chloride diffusivity determination of cement-based materials. *Journal of The Chinese Ceramic Society.* 2012, 40:651–656.

Xia J, Li L-y. Numerical simulation of ionic transport in cement paste under the action of externally applied electric field. *Construction and Building Materials.* 2013, 39:51–59.

Xu A. *The Structure and Some Physical and Properties of Cement Mortar with Fly Ash.* Chalmers University of Technology, Goteborg, Sweden; 1990.

Xu J, Zhang C, Jiang L, et al. Releases of bound chlorides from chloride-admixed plain and blended cement pastes subjected to sulfate attacks. *Construction and building Materials.* 2013, 45:53–59.

Xu Y. The influence of sulphates on chloride binding and pore solution chemistry. *Cement and Concrete Research.* 1997, 27:1841–1850.

Yang Z, Gao Y, Mu S, et al. Improving the chloride binding capacity of cement paste by adding nano-Al2O3. *Construction and Building Materials.* 2019, 195:415–422.

Ye H, Jin X, Fu C, et al. Chloride penetration in concrete exposed to cyclic drying-wetting and carbonation. *Construction and Building Materials.* 2016, 112:457–463.

Yong R. Swelling pressures of sodium montmorillonite at depressed temperatures. *Clays and Clay Minerals.* 1962, 11:268–281.

Yu H, Weng Z, Sun W, Chen H, Zhang J. Influences of slag content on chlorine ion binding capacity of concrete. Journal-*Chinese Ceramic Society.* 2007, 35:801.

Yuan Q. *Fundamental Studies on Test Methods for the Transport of Chloride Ions in Cementitious Materials.* Ghent University, Belgium; 2009.

Yuan Q, Shi C, De Schutter G, Audenaert K, Deng D. Chloride binding of cement-based materials subjected to external chloride environment—A review. *Construction and Building Materials.* 2009, 23:1–13.

Zhang M-H, Gjørv OE. Effect of silica fume on pore structure and chloride diffusivity of low parosity cement pastes. *Cement and concrete Research.* 1991, 21:1006–1014.

Zheng L, Jones MR, Song Z. Concrete pore structure and performance changes due to the electrical chloride penetration and extraction. *Journal of Sustainable Cement-Based Materials.* 2016, 5(1–2):76–90.

Zhu Q, Jiang L, Chen Y, Xu J, Mo L. Effect of chloride salt type on chloride binding behavior of concrete. *Construction and Building Materials.* 2012, 37:512–517.

Zhu X, Zi G, Cao Z, et al. Combined effect of carbonation and chloride ingress in concrete. *Construction and Building Materials.* 2016, 110:369–380.

Zibara H, Hooton R, Thomas M, Stanish K. Influence of the C/S and C/A ratios of hydration products on the chloride ion binding capacity of lime-SF and lime-MK mixtures. *Cement and Concrete Research.* 2008, 38:422–426.

Chapter 5

Testing Methods for Chlorides Transport in Cement-Based Materials

5.1 INTRODUCTION

It is well known that chloride ion penetrates into concrete at a very slow rate in practice or in application. It takes a very long time to reproduce the real engineering conditions, and this is unacceptable for experiments in the lab and tests in the field. Fast results, however. are always desired in engineering applications. Various techniques were used to accelerate chloride transports in concrete to shorten the test duration and obtain fast results. Accordingly, different theoretical bases were applied to assess the chloride resistance of concrete.

During the past several decades, a great amount of effort has been devoted to evaluating the transport of chlorides in concrete, due to its significance. Thus, many methods have been proposed and developed to measure the chloride transport in concrete. However, there is no one strict way to categorize the existing methods. Streicher and Alexander (1994) and Stanish et al. (2001) have given reviews on the existing methods. Shi et al. (2007) also gave a review on the existing methods, in which the advantages and disadvantages were discussed. According to Shi et al. (2007), based on whether chloride concentration in concrete changes with time during the test duration, they can be divided into steady and non-steady state test methods.

(1) Steady-state test method

Steady state refers to a state that the chloride ion concentration at each point in the sample already reaches equilibrium, and doesn't change with time. In other words, concrete is saturated with chlorides, and the quantity of chloride penetration into the sample equals to that moves out of the sample.

(2) Non-steady-state test method

In contrast, chloride ion concentrations still change with time at some points of the sample under non-steady-state conditions.

Based on the experimental conditions and principles, they can be classified into diffusion test methods, accelerated migration test methods, electrical conduction methods, and some other methods (Shi et al. 2007).

Actually, the main two purposes for the measurement of chloride transport in concrete are ranking the chloride resistance of concrete or predicting service life of the concrete structure subjected to chloride environments.

Some test methods can only fulfill the first purpose, like ASTM C 1202 (i.e., electrical indication of concrete's ability to resist chloride ion penetration) in which only the charge passed through concrete specimens within six hours is measured. This cannot be directly used for the prediction of the service life of the concrete structure. This type of test is more suitable to be a quality control tool. In contrast, the diffusion coefficient of concrete can be measured by some test methods. This type of test methods can fulfill both two purposes. Obviously, the latter one is more desirable.

In this chapter, a classification of the existing methods based on the theoretical bases behind the calculation of diffusion coefficient is presented. The disadvantages and advantages of these methods are discussed, together with the detailed procedure of testing methods. The testing results obtained from different methods are also discussed in detail.

In addition to the characterization method of chloride transport in concrete, some testing methods on chloride contents and its sampling method should be discussed first, such as water-soluble chloride, total chloride and free chloride.

5.2 SOME CHLORIDE-RELATED TESTS

5.2.1 Chloride Profile

Chloride profile in the specimen along the transport direction is often desired in many applications, such as determination of chloride diffusion coefficient and evaluation of corrosion risk of reinforced concrete structure.

Since the chloride penetration depth is often small, in order to obtain enough points for the profile, the thickness of sampling has to be very thin. The chloride profile is often measured on specimens by successively grinding off in layers (0.5-1mm) parallel to the exposed surface. Therefore, special instruments are needed for sampling. Germann Corp. developed an instrument, named Profile Grinder 1100. The increment is controlled by turning the grinding housing against the handle cover. There are four engraved red spots on the top of the casing, equally spaced 90 degrees. 90 degrees corresponds to 0.5mm depth increment. The handle cover has also one such engraved red spot. By positioning the spot on the handle cover relative to the spot on the grinding housing, the depth increment can be adjusted. If, for example, 1 mm is needed, one of the red spots of the grinding housing

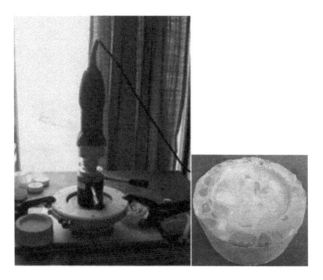

Figure 5.1 Profile grinding machine and specimen after the process of grinding.

is turned 180 degrees relative to the red mark on the handle cover. The specimen is placed centrally on the bottom plate which is firmly secured to a table with two screw clamps, as shown in Figure 5.1. Adjust the position of the diamond bit to the top surface of the specimen, followed by adjusting the red spots on grinding housing and handle cover, and then the grinding can be started. The grinding area is 73 mm in diameter, as shown in Figure 5.1. Exact depth increments are adjustable, from 0.5 mm to 2.0 mm. The depth increments are accurate within 2% and the variation is less than 1%. The maximum grinding depth is 40 mm. The powder produced was first collected with a spoon, and then cleaned with a vacuum cleaner. For every depth increment of 0.5 mm, approximately 5 grams of powder is available for both total and water-soluble chloride analysis.

A more advanced lathe for chloride profile was developed by Shi (2010) and was already commercially available in China, as shown in Figure 5.2. The sampling thickness and the collection of powder can be controlled by the machine automatically. The sampling thickness can be as small as 0.1 mm. This machine greatly save labor for obtaining the powder at different depth, and accurate results can be achieved.

5.2.2 Chloride Analysis

Chlorides presenting in concrete include the chlorides in pore solution and the chlorides bound to cement hydration products. The equilibrium can be influenced by many factors, such as temperature, pH value, external forces,

Figure 5.2 Automatic lathe for chloride profile sampling.

etc. How to define free chloride is a very important question. The following are some options for the description of free chloride (Nilsson 2002).

- Free to move
- Free to be leached out
- Free to corrode steel

The first term refers to the chlorides in the pore solution, which can move freely. This type of chloride is very important in modeling chloride transport in concrete, because only this type of chloride contributes to the chemical potentials. The second term represents that all chlorides may be released into solvents, mainly distilled water, which will be the sum of the chlorides in pore solution and some loosely bound chlorides released by solvent. Alkaline solution was also used as solvent by Castellote et al. (1999a). The third type of chloride is the one of interest. Not only the chlorides in pore solution but also the bound chlorides could corrode steel which may be released into pore solution due to the change in tempera-ture, chemistry of pore solution, the effect of carbonation, etc. However, to today's knowledge, it is difficult to precisely distinguish and quantify the chlorides which are harmful to steel because of its complexity. Roughly, the first type of chloride is regarded as to be harmful to steel. Therefore, in many researches, free chloride is defined as the chlorides present in pore solution of cement-based materials.

The concentration of chloride in pore solution cannot be determined directly. Pore solution expression method is probably the most used and accurate method to determine free chloride concentration in the pore solu-tion (as shown in Figure 5.3), although Glass et al. (1996) pointed out that high pressure during squeezing out the pore solution might release some loosely bound chlorides. There are no other techniques which are better than pore expression method in determining the chemistry of pore solution.

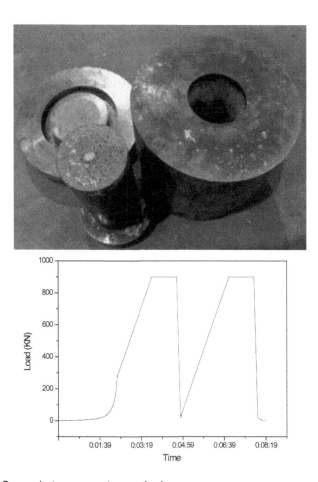

Figure 5.3 Pore solution expression method.

Thus, it is often taken as a reference method. Nevertheless, pore expression method has the following drawbacks: (1) it needs special equipment; (2) it is not easy in manipulation; (3) it is very hard to get pore solution from concrete with low water-to-binder ratio; (4) it is almost impossible to determine chloride content in a very thin layer sample, which is required in the curve fitting method. A method is needed to replace pore expression method. Liquid extraction methods are the good options, among which water extraction is the most used one. Castellote et al. (1999a) proposed to extract chloride by using alkaline solution. A solution to solid ratio of 2/3, particle size of 2.5-3.5 mm, and contact time of 24 hours in inert environment were suggested. Castellote et al. calibrated the alkaline extraction method with pore expression method and found that this method is good enough to estimate the results of pore expression method. It is worth to mention that water-soluble chloride is more often used to estimate free chloride, and more practical. Thus, it is necessary to establish the relationship between free chloride and water-soluble chloride.

Some methods have been standardized to determine chloride content in concrete.

- China: JGJ/T 322-2013, Technical specification for test of chloride ion content in concrete
- Europe: NT build 208-96, Concrete, hardened: chloride content by Volhard titrate
- America: ASTM C 1218 C-99, Standard test method for water-soluble chloride in mortar and concrete

The above-mentioned standards adopt the same principle, and may have some differences in detail. The testing procedures are described briefly as follows:

5.2.2.1 Determination of Total Chloride

The procedure used to analyze the total chloride content is described as follows:

- Dry the powder (passed through 0.1 mm sieve) at 105°C to constant mass and cool it down to room temperature.
- Take 5 g sample and place it in a 150ml beaker.
- Add about 20 ml of distilled water and shake the bottle to achieve separation of the particles. Add about 10 ml concentrated nitric acid, shake the bottle, add about 50 ml hot distilled water and shake again. Let the mixture cool for about one hour until it reaches ambient temperature.
- Wait until the solution becomes cool, then gently pour solution onto the filter paper, and then rinse the beaker onto the filter paper with distilled water.

- The total solution is adjusted to 100ml, and then 10ml solution is pipetted for the determination of the concentration of chloride.

Chloride concentration can be analyzed by means of potentiometric titration or Volhard titrate. Potentiometric titration gives better precision and accuracy. A Metrohm MET 702 automatic titrator can be used for this purpose, as shown in Figure 5.4. 0.01mol/l silver nitrate solution was used as titration solution. A magnetic stirrer was used for dispensing solution during titration, and the titrator can calculate the chloride concentration according to the mv-volume curve automatically. The chloride content is given by:

$$c_t(\%) = \frac{10 \times 100 \times V \times 35.45 \times 0.01}{1000 \times 2} \tag{5.1}$$

where c_t (%) is the total chloride content by the mass of sample, V is the titrated silver nitrate solution, ml.

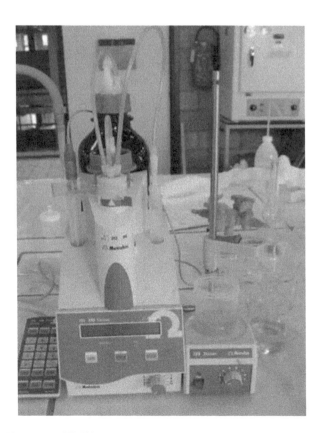

Figure 5.4 Metrohm MET 702 automatic titrator.

5.2.2.2 Determination of Water-Soluble Chloride

The method used to determine the water-soluble chloride content is described as follows:

- Dry the powder (passed through 0.85 mm sieve) at 105°C to constant mass and cool it down to room temperature.
- Weigh approximately 10 g sample to the nearest 0.01 g and place it into a 250-mL beaker.
- Add 50 mL of reagent water into the beaker, bring to a boil and boil for 5 min. Allow to stand 24 h.
- Filter by gravity or suction through a fine-texture. Transfer the filtrate to a 250-mL beaker. Add 3 mL of (1:1) nitric acid to the filtrate.
- Cover the beaker with a watch glass and allow to stand for 1 to 2 min. Heat the covered beaker rapidly to boiling.
- 1ml solution is pipetted for the determination of the concentration of chloride.

The water-soluble chloride can be calculated as:

$$c(\%) = \frac{10 \times 100 \times V \times 35.45 \times 0.01}{1000 \times 2.5} \tag{5.2}$$

where $c\%$ is the water-soluble chloride content by the mass of sample, V is the titrated silver nitrate solution, ml.

The unit of the determined water-soluble chloride is percentage by the mass of sample. However, the free chloride in the unit of mol/l is more of interest. In combination with the water accessible porosity of sample, the unit of chloride can be transformed from % to mol/l:

$$c(\text{mol/l}) = \frac{1000 \times c\%}{35.45 \times \omega_w\%} \tag{5.3}$$

where $c\%$ is the chloride content by the mass of concrete, $\omega_w\%$ is the water content of concrete by the mass of concrete, $c(\text{mol/l})$ is the chloride concentration in pore solution.

With the water content by volume, the chloride in unit of mol/l can also be obtained by,

$$c(\text{mol/l}) = \frac{\rho_{dry} \times c\%}{35.45 \times \omega\%} \tag{5.4}$$

where ρ_{dry} is the dry density of concrete, $\omega\%$ is the water content of concrete by the volume of concrete.

5.2.2.3 Relationship between Water-Soluble and Total Chlorides

It is reported that water-soluble chloride is higher than free chloride. Otsuki et al. (1992) proposed a linear relationship between water-soluble and free chlorides: $C_w = 1.83\,C_f$ or $C_f = 0.546\,C_w$ (C_W is the value obtained by the stirring extraction method; C_f is the value obtained by the pore expression method). Tetsuya et al. (2008) used power relationship to describe the relationship between water-soluble chloride and free chloride. Using the data from Haque and Kayyali (1995), the relationship between free chloride and water-soluble chloride are plotted in Figure 5.5. Results in Figure 5.5 indicate that both linear and power equations can be used to approximately describe the relationship.

Actually, the amount of chloride extracted by water is influenced by many factors, such as water to solid ratio, temperature, size of particle, leaching time, leaching solvent, etc (Vladimir 2000). Therefore, the relationships between water exaction method and pore expression method reached by different authors normally have no universality. The relationship between water extraction method and pore expression method was established. Castellote et al. (1999a) studied the effect of contact time on the leaching method and found that 24 hours was the optimum time to reach the equilibrium between the pore solution in particles with the size of 2.5–3.5 mm and the exposure solution.

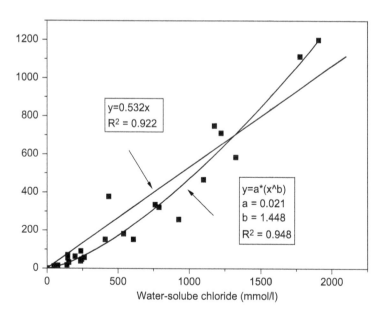

Figure 5.5 Relationship between water-soluble chloride and free chloride. (From Haque 1995.)

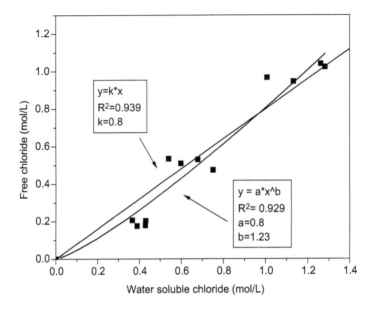

Figure 5.6 Relationship between free chloride and water-soluble chloride obtained by Yuan (2009).

Yuan (2009) studied the relationship between water-soluble chlorides and free chlorides of cement paste sample. Power and linear relationships were used to correlate water-soluble chloride and free chloride, as shown in Figure 5.6. Both relationships fit experiment data quite well. The correlation coefficient of linear fitting is higher than that of non-linear fitting. Nevertheless, it is hard to say which relationship is better based on the limited data. For simplifying reasons, a linear relationship was used to convert water-soluble chloride into free chloride (Yuan 2009),

$$c = 0.8 \times c_w \tag{5.5}$$

where c is the calculated free chloride in pore solution (mol/l), c_w is the determined water-soluble chloride (mol/l).

The factor of 0.8 in Equation 5.5 is quite different from Haque's 0.532 (1995). Different experimental conditions may partly account for this. In addition, the water-soluble chloride used by Haque and Kayyali (1995) is different from Yuan's study (2009). Haque and Kayyali (1995) directly used the chloride concentration in solvent (distilled water) as the concentration water-soluble chloride. In contrast, water-soluble chloride used in (Yuan 2009) is based on the chlorides released into water and evaporable water content of concrete. The application of Equation 5.5 still needs more data for validation.

5.3 TESTING METHODS FOR CHLORIDE TRANSPORT IN CONCRETE

5.3.1 Brief Overview of Testing Methods

As stated above, since diffusion is a very slow process, various techniques, which mainly include DC, AC, and high pressure, have been used to accelerate the process; thus, fast testing results can be obtained. Among the accelerated techniques, DC is the most used one.

The testing methods are based on various theoretical bases. Table 5.1 gives a summary of the test methods. The disadvantages and advantages of the methods are briefly presented in Table 5.1. Based on the theoretical bases, the test methods can be grouped into six categories:

- Fick's first law
- Fick's second law
- Nernst–Planck equation
- Nernst–Einstein equation
- Formation factor
- Other

As can be seen from Table 5.1, the Nernst–Planck equation is the most often used theory in the calculation of migration coefficient.

Different terminologies on chloride transport coefficient appeared in literature, such as diffusion/migration coefficients, effective diffusion/migration coefficients, steady-state diffusion/migration coefficients, non-steady-state diffusion/migration coefficients and apparent diffusion coefficient, etc. These terms are little confusing. It is necessary to make them clear. Migration coefficient actually is diffusion coefficient determined from electrically accelerated test, not natural diffusion test. The aim of electrically accelerated tests is to estimate diffusion coefficient. In some literature, the coefficients determined from electrically accelerated tests were just called diffusion coefficients, while in other literature, they were called migration coefficients. Effective diffusion/migration coefficients are equal to steady-state diffusion/migration coefficients. Their relationship is shown in Figure 5.7.

To make the terminology clear, the diffusion coefficients determined from electrically accelerated tests are called non-steady-state (D_{nssm}) or steady-state (D_{ssm}) migration coefficients, diffusion coefficients obtained from natural diffusion tests are called non-steady-state (D_{nssd}) or steady-state (D_{ssd}) diffusion coefficients.

The following sections give detailed description of the testing method for chloride transport in concrete.

Table 5.1 Summary of Test Methods for Chloride Transport in Concrete

Theoretical Base	Test Method	Measurement	Test Duration	Remark	Reference
Fick's first law	Steady-state diffusion test	Chloride flux	Several months	Long duration, not easy to perform	Page (1981)
Fick's second law	NT build 443	Chloride profile	>35days	Close to reality, free chloride should be used and not easy to perform.	NT build 443 (1995) ASTM C1556-11a (2016)
	Short-term immersion test	Chloride concentration reduction	2 weeks	Easy to perform, need more validations.	Park (2014)
Nernst–Planck equation	NT build 355	Chloride flux	Several weeks	Single species theory.	NT build 355
	Truc's method	Chloride flux	Several days	Single species theory, and chloride flux from upstream may depend on binding.	Truc (2000a)
	NT build 492	Penetration depth	24-72 hours	Single species theory, and colorimetric method is not accurate.	NT build 492
	Breakthrough time method	Breakthrough time	Several weeks	Theory is not clear, and too many definition of breakthrough time.	Halamickova (1995)
	Andrade and Castellote's method	Various theory	Several days	Many methods were proposed by Andrade et al.	Andrade (2000a)
	Samson's method	Current	120 hours	Soundest theoretical base, but too complicated.	Samson (2003)
	Friedmann's method	Current	Several weeks	Well thought method, only for steady-state test.	Friedmann (2004)
Nernst–Einstein equation	Lu's method	Resistivity	Several minutes	Concerns on the correction factor of f and technique for the saturation of concrete.	Lu (1998)

(Continued)

Table 5.1 (Continued) Summary of Test Methods for Chloride Transport in Concrete

Theoretical Base	Test Method	Measurement	Test Duration	Remark	Reference
Formation factor	Formation factor Method	Resistivity	Several minutes	Concerns on the technique for the saturation of concrete.	Streicher and Alexander (1995) Azad et al. (2019)
Other	ASTM C1202 or AASHTO T227	Charge passed	6 hours	All conductive ions are responsible for the charge passed.	ASTM C1202-2005
	AASHTO T 259 90day ponding test	Chloride profile	90days	Two mechanisms involved, the effect of individual is not clear.	AASHTO T 259
	Water pressure method	Penetration depth	Several weeks	Needs special equipment.	Freeze (1979), (Stanish et al. (2001))
	AC impendence method	Impedance	Several minutes	Good for measurement of conductivity of concrete.	Shi (1999)

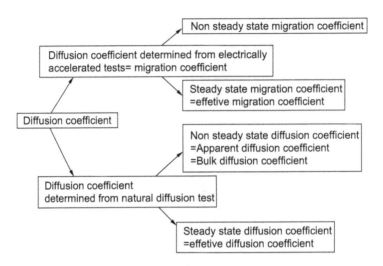

Figure 5.7 Terms related with diffusion coefficient.

5.3.2 Fick's First Law

The Fick's first and second laws were proposed by analogy to the Fourier's laws of heat conduction (Crank 1975). As such the implicit assumptions of the Fourier's Laws have to be fulfilled whenever Fick's laws are to be used. In the case of ion transport in cement-based materials, the assumptions that have to be fulfilled include: (1) ions move independently of each other; (2) concentration gradient is the only driving force; (3) ions have very weak or no interaction with matrix (Chatterji 1995). Obviously, the first and second assumptions are not fulfilled. Pore solutions of cement-based materials are normally filled with various ions. The interactions between different ions are very strong. Ions move under not only concentration gradient but also the electrical field created by the movements of other species. The third assumption is also not fulfilled; ions may bind to cement hydration products, and the ions may interact with solid surface due to the effect of electrical double layers.

Despite the fact that all the assumptions are not fulfilled, Fick's first law is still applied to study chloride transport in cement-based materials under steady-state conditions (Page et al. 1981; Byfors 1987; Hansson and Sorensen 1987; Arsenault 1995). Figure 5.8 illustrates the typical diffusion cell.

The two solution chambers beside the specimen often contain chloride-free solution and chloride solution. When chlorides from the upstream chamber reach the downstream chamber, the chloride concentration in the downstream chamber is then determined periodically. Steady state is believed to be reached when the chloride flux is stable. The steady-state

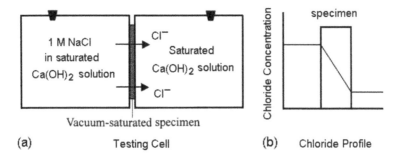

(a)   Testing Cell   (b)   Chloride Profile

Figure 5.8 Diffusion testing cell and chloride ion concentration profile.

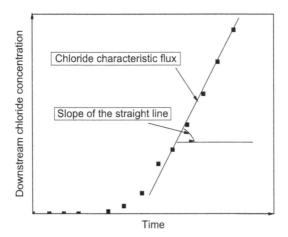

Figure 5.9 Evolution of chloride concentration in chloride-free cell during diffusion test.

chloride flux can be characterized by the slop of the straight line of chloride concentration against time, as shown in Figure 5.9. The diffusion coefficient can then be calculated,

$$J = D_{ssd} \frac{c_0 - c_1}{L} \tag{5.6}$$

here c_0 is the chloride ion concentration in the upstream cell, c_1 is the chloride ion concentration in the downstream cell, and L is the thickness of the specimen, and D_{ssd} is the steady-state diffusion coefficient.

In addition to the fact that assumptions cannot be fulfilled, the main drawback of this method is that the test takes months to complete since the diffusion of chloride ion through hardened cement pastes or concrete is a very slow process.

5.3.3 Fick's Second Law

5.3.3.1 NT Build 443

Despite the theoretical errors mentioned above, Fick's second law is also widely used to measure the chloride transport in cement-based materials under non-steady-state conditions. AEC laboratory in Denmark first developed the bulk diffusion test (AEC Laboratory 1991), which is based on Fick's second law. Afterwards, it became the first formally standardized version of the bulk diffusion test method, NT Build 443-94. The testing setup is shown in Figure 5.10. The testing specimen should have a minimum maturity corresponding to curing for 28 days at 20°C and needs to be saturated with limewater before testing to avoid absorption. This means diffusion is the only transport mechanism in this test. All faces of the specimen are covered, but one face is left uncovered and is exposed to a 2.8 mol/l NaCl solution, as shown in Figure 5.10. After a minimum of 35 days exposed to the NaCl solution, the concrete specimen is ground off repeatedly parallel to the exposed surface at depth increments on the order of 0.5–1 mm. The acid-soluble chloride content of the samples is then determined.

The values of c_s and D_{app} are determined by fitting Equation 5.7, which is an error function solution to the Fick's second law, to the measured chloride profile by means of a non-linear regression analysis in accordance with the method of least squares fit, as shown in Figure 5.11. The first point of the profile determined from the sawn face is omitted in the regression analysis. The other points are weighted equally.

$$c(x,t) = c_s - (c_s - c_i).\mathrm{erf}\left(x / \sqrt{4D_{app}t}\right) \tag{5.7}$$

where $c(x,t)$ is chloride concentration measured at the depth x at the exposure time t, c_s is the surface chloride concentration, c_i is the initial chloride concentration measured on the concrete slice, x is the depth below the exposed surface (to the middle of a layer), D_{app} is the apparent (bulk)

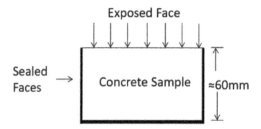

Figure 5.10 Illustration of NT build 443 testing setup.

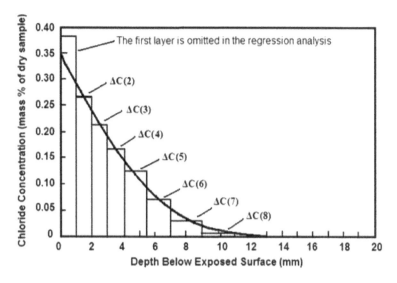

Figure 5.11 Regression analysis of non-steady-state diffusion test results.

chloride diffusion coefficient, also called non-steady-state diffusion coefficient, and t is the exposure time.

This test measures non-steady-state diffusion coefficient of chloride ion in saturated concrete. It resembles many cases in real concrete structures which are under seawater. However, it also takes long time to conduct the test. For low quality concretes, the minimum exposure period is 35 days. For higher quality concretes, however, this period must be extended to 90 days or even longer. Furthermore, this test is laborious and needs expensive equipment for chemical analysis.

5.3.3.2 Short-Term Immersion Test

Park et al. (2014) proposed a method to obtain the non-steady-state diffusion coefficient in the short term, as shown in Figure 5.12. In such a test, concrete specimen is ponded in chloride solution, and the chloride concentration is measured at 0, 1, 2, 3, 4, 5, 7, 10, and 14 days. To minimize the total volume change of the source solution, the sampling size was chosen to be 0.5 ml, which is as small as possible.

Based on Fick's second law, Park et al. (2014) obtained analytical solution for evolution of chloride concentration of ponding solution with time, as follows:

$$C_{\text{source}}(t) = c_{\text{source}}(0) \cdot \exp\left[D_{ST} \cdot \frac{t}{h^2}\right] \cdot \text{erfc}\left[\left(D_{ST} \cdot \frac{t}{h^2}\right)^{\frac{1}{2}}\right] \tag{5.8}$$

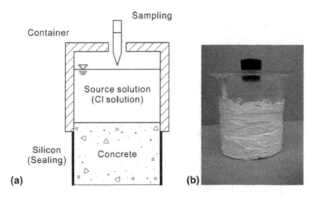

Figure 5.12 (a) Test setup for short-term immersion test and (b) photo of the test setup. (From Park et al. 2014.)

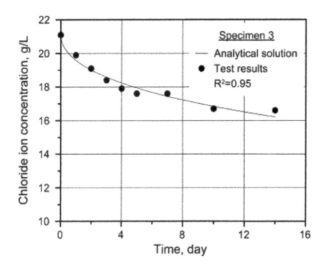

Figure 5.13 Calculation of diffusion coefficient in 21 g/L source solution for mortar. (From Park et al. 2014.)

The diffusion coefficient can be determined by curve-fitting to minimize the difference between the measured chloride ion concentrations and the profiles calculated by the analytical solution. The least squares method was employed to obtain a diffusion coefficient, i.e., a curve minimizing the difference between the measurements and the calculations, as shown in Figure 5.13.

5.3.4 Nernst–Planck Equation

Natural diffusion tests take a long time to obtain results. The movement of chloride ion can be accelerated through the use of an electrical field. Many electrically accelerated tests have been proposed or developed to measure the diffusion of chloride ion in concrete. By applying electrical field, the results can be obtained within a reasonable time.

If considering concrete as a solid electrolyte, the quantification of ionic movement in electrolytes can be solved in steady-state conditions by using Nernst–Planck equation (Bockris 1982; Andrade 1993):

$$J(x) = -\left[D\frac{\partial c(x)}{\partial x} + \frac{ZF}{RT} Dc\frac{\partial E(x,t)}{\partial x} + cV(x) \right] \tag{5.9}$$

where D is the diffusion coefficient of chloride in concrete, c is the chloride concentration in pore solution, $V(x)$ is the advection term.

Each term on the right-hand side of Equation 5.9 corresponds to different mechanisms. The first term, often called the diffusion term or Fick's first law, describes the movement of ions under the effect of a concentration gradient. The second is the movement of ions under electrical field, which may be the combination of membrane potential and external electrical field. The third term is the advection term; no pressure gradient is considered. Thus, this term is omitted from the equation. Equation 5.9 then becomes:

$$J(x) = -\left[D\frac{\partial c(x)}{\partial x} + \frac{ZF}{RT} Dc\frac{\partial E(x,t)}{\partial x} \right] \tag{5.10}$$

Equation 5.10 has been widely used as theoretical base for the calculation of diffusion coefficient determined from electrically accelerated tests under both steady and non-steady state (Andrade et al. 1993, 1994, 1996; Zhang et al. 1994; Truc et al. 2000; Samson et al. 2003; Friedmann et al. 2004; Krabbenhøft et al. 2008). A typical testing setup consists of two chamber cells and a concrete sample as the division between the two chambers, as shown in Figure 5.14. The concrete sample can be any size but is usually a disk of 100 mm diameter and length from 15 to 50 mm. A sufficient length is required for the specimen to avoid aggregate interface influences.

5.3.4.1 NT Build 355

Andrade (1993) solved Equation 5.9 based on the following assumptions:

- The ionic mobility is much higher in the solutions than in the concrete disc and therefore the "distance" of the experiment is the disc thickness.
- Advection does not exist in the tested concrete specimens (i.e., no pressure or moisture gradients).
- Diffusion portion is negligible compared to the effect of electrical migration when a sufficiently strong applied voltage (at least 10 V) is applied.
- Chloride concentration is constant in the upstream cell.
- The electrical field along the concrete disc follows a linear decay.

Figure 5.14 Migration processes and reactions happening in an accelerated Cl- migration testing cell.

Then Equation 5.9 becomes:

$$-J(x) = \frac{ZF}{RT} D_{eff} c \frac{\partial E(x)}{\partial x}$$ (5.11)

and

$$D_{eff} = \frac{RTLV_1}{ZFE\gamma c_0 A} \cdot \frac{\Delta c_1}{\Delta t}$$ (5.12)

where L is the thickness of concrete specimen, c_0 is the upstream chloride concentration, Δc_1 is the change in downstream chloride concentration, V_1 is the downstream volume, A is the area of the side of specimen exposed to chloride solution, Δt is the time increment, γ is the activity coefficient, and Z is the valence of chloride.

This method was standardized as NT build 355 in 1997. It is worth it to mention that the activity coefficient is not presented in the equation of NT Build 355, as follows:

$$D_{eff} = \frac{RTLV_1}{ZFEc_0 A} \cdot \frac{\Delta c_1}{\Delta t}$$ (5.13)

This implicitly assumes that the activity coefficient equals to 1 under the action of strong electrical field. The main drawbacks of this method include:

- It takes quite long time to reach steady state, especially for high performance concrete.

- Chloride ions may turn into chorine in the downstream cell, and disappear as gas. This reaction is difficult to control and to quantify.
- The electrical potential between the two electrodes, instead of the real electrical potential across the specimen, is used in the calculation. In other words, the potential drop between the electrodes and specimen surface is not taken into account.
- It is based on a single species theory. The interactions between chloride ions and other species are not taken into account.

To overcome the fourth drawback, Zhang et al. (1995) introduced a correction factor β_0 for the calculation of chloride migration coefficient of concrete:

$$D_{eff} = \beta_0 \frac{RT}{ZFE} \cdot \frac{LV_1}{c_0 A} \cdot \frac{\Delta c_1}{\Delta t} \tag{5.14}$$

The correction factor is dependent upon the concentration and temperature of the solution. Table 5.2 lists some correction factors for ionic interaction from NaCl solution at different temperatures and different concentrations (Zhang et al. 1995).

5.3.4.2 Upstream Method (Truc's Method)

It is time consuming to perform NT build 355 test, and the test is not easy to perform either. Moreover, chemical reactions can occur at the anode. Those reactions are difficult to control and to quantify. The value of the migration coefficient is therefore determined with a large uncertainty. Under steady-state conditions, the flux of chloride ions penetrating into the specimen from the upstream cell is equal to that moving out of specimen into the downstream cell. Truc et al. (2000) proposed a method to use the flux of chloride ions through the upstream side of the concrete specimen J_{up} to calculate the diffusion coefficient under steady-state conditions:

$$D_{eff} = \frac{J_1 RTL}{c_0 FE} = \frac{RTLV_0}{ZFEc_0 A} \cdot \frac{\Delta c_0}{\Delta t} \tag{5.15}$$

Table 5.2 Correction Factor for Ionic Interaction from NaCl Solution at Different Temperatures and Different Concentrations

NaCl Concentration (Mol/L)	Correction Factor at Different Temperatures					
	20°C	21°C	22°C	23°C	24°C	25°C
0.1	1.06	1.06	1.06	1.07	1.07	1.08
0.2	1.16	1.17	1.18	1.19	1.19	1.20
0.3	1.26	1.27	1.29	1.29	1.30	1.30
0.4	1.36	1.37	1.39	1.40	1.42	1.43
0.5	1.46	1.48	1.50	1.52	1.53	1.55

Zhang et al. (1995).

where J_1 is the chloride ion flux through the upstream side of the sample, c_0 is chloride concentration in the upstream cell, and $\triangle c_0$ is the change in upstream chloride concentration.

This method can overcome many drawbacks of NT build 355. It assumes that the upstream flux is constant at the very beginning and therefore independent of chloride binding. The upstream diffusion coefficient $D_{eff,up}$ is equal to the downstream diffusion coefficient $D_{eff,down}$, even when the steady-state in downstream cell is not achieved yet. For an ordinary concrete, two or three days can give satisfactory results (Truc et al. 2001). This greatly shortens the testing time and simplifies the testing procedure. It may also be possible to use this method to measure the diffusion coefficient of concrete already contaminated by chloride ions. However, the assumption that the upstream flux is constant at the very beginning, and independent on chloride binding, is highly questionable.

5.3.4.3 NT Build 492

Tang and Nilsson (1992b) solved Equation 5.9 under non-steady migration conditions by assuming a semi-infinite diffusion with the following initial and boundary conditions:

$$C = C_0, x = 0 \ t > 0$$

$$C = 0, \ x > 0 \qquad t = 0$$

$$C = 0, \quad x \rightarrow \infty t = t_M$$

where t_M is a finite large number.

Then the analytical solution for Equation 5.8 is as follows:

$$c = \frac{c_0}{2}\left[e^{ax}\text{erfc}\left(\frac{x + aD_{nssm}t}{2\sqrt{D_{nssm}t}}\right) + \text{erfc}\left(\frac{x - aD_{nssm}t}{2\sqrt{D_{nssm}t}}\right)\right] \tag{5.16}$$

where $a = zFE / RTL$, erfc is the complement to the error function erf, and D_{nssm} is the non-steady-state migration coefficient. When the electrical field E/L is large enough and the penetration depth x_d is sufficient ($x_d > aDt$), the second term on the right-hand side of Equation 5.16 approaches to zero and can be ignored. Thus, the above equation becomes:

$$c_d = \frac{c_0}{2}\text{erfc}\left(\frac{x - \alpha D_{nssm}t}{2\sqrt{D_{nssm}t}}\right) \tag{5.17}$$

After some mathematical transformations, it turns into,

$$D_{nssm} = \frac{RTL}{FE} \cdot \frac{x_d - \alpha\sqrt{x_d}}{t} \tag{5.18}$$

with

$$\alpha = 2\sqrt{\frac{RTL}{FE}}\,\text{erf}^{-1}\left(1-\frac{2c_d}{c_0}\right)$$ (5.19)

where x_d is the average chloride ion penetration depth measured by spraying 0.1 mol/l silver nitrate solution, c_d is the chloride concentration at which the color changes (0.07 mol/l for Portland cement concrete), E is the applied voltage, T is the average value of the initial and final temperatures in the anolyte solution, L is the thickness of the specimen, and t is the test duration.

This method has been standardized by NordTest as NT Build 492 in 1999. The test method recommends varied applied voltages and testing durations based on measured initial currents through tested specimens so to avoid significant heating during the testing and to obtain a reasonable chloride penetration depth. At the end of the testing, the specimens were axially split into two pieces and 0.1 mol/l silver nitrate solution is used to spray on the freshly split section to determine the chloride ion penetration depth.

It is worth it to mention that due to the potential drop between electrodes and the surface of specimen, the real potential across the specimen is lower than the potential between two electrodes which is used in the calculation of migration coefficient. Systematic studies (McGrath and Hooton 1996) showed that the real potential across specimen is 1.5–2 V lower than the electrode potential. Therefore, 2 volts are subtracted from the potential between two electrodes.

10% (approximately 2 mol/l) NaCl solution is specified in NT build 492; thus,

$$\text{erf}^{-1}\left(1-\frac{2c_d}{c_0}\right) = \text{erf}^{-1}\left(1-\frac{2\times0.07}{2}\right) = 1.28$$

NT build 492 gives a more practical equation for the calculation of D_{nssm}, as follows:

$$D_{nssm} = \frac{0.0239(273+T)L}{(E-2)t}\left(x_d - 0.0238\sqrt{\frac{(273+T)Lx_d}{E-2}}\right)$$ (5.20)

This method has the following advantages:

- Short testing period
- Simple measurement
- Simple calculation
- No strict sealing requirement
- Clear theoretical bases

In Streicher's review paper (Streicher and Alexander 1994), Tang's method seems to be "the most suitable of all the rapid chloride tests reviewed, on the basis of simplicity, duration of test, theoretical basis and versatility."

However, this method also has the following drawbacks:

- The measurement of chloride penetration depth by using colorimetric method is by visual judgment. It is not easy to measure accurately.
- Chloride concentration at the color change boundary, i.e. C_d, is dependent upon the alkalinity of the concrete.
- It is based on single species theory. The theory considers that the chloride ions move independently, without interaction with other species existing in pore solution of concrete.
- The chloride front given by Equation 5.16 is very sharp, and it is not confirmed by many experimental data, as shown in Figure 5.15. This disagreement between theory and experimental data motivated Stanish et al. (2004) to formulate a new model for chloride transport in concrete. On the contrary, Voinitchi's results (Voinitchi et al. 2008) showed that total chloride profile exhibits a plateau on the cathodic side, combined with a classical diffusion profile towards anodic side, as shown in Figure 5.15. This fits exactly Tang's theoretical figure. It is worth it to mention that the chloride concentration in Tang's theoretical model is free chloride. Since measuring free chloride in very thin thickness is not easy, most researchers only measure total chloride at each depth. How the free chloride profile of a specimen after migration test looks is still uncertain. Spiesz et al. (2013) explained that the discrepancy between the theoretical abrupt chloride concentration profile and the experimental gradual profiles can be attributed to the fact that at higher

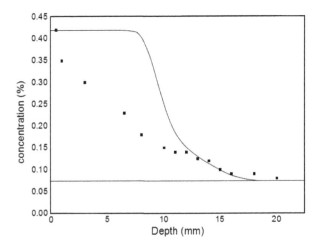

Figure 5.15 Experimental results (dots) and predicted profile (line) by Equation 5.16. (From Stanish et al. 2004.)

chloride concentrations the chloride binding capacity is increased. This in turn reduces the apparent chloride migration coefficient in the layers of concrete with increased chloride concentrations and causes a higher accumulation of chlorides in these layers compared to the layers with lower c (closer to the chloride penetration front).

Gao et al. (2017) verified the influence of electrical conductivity of pore solution on the chloride migration coefficient measure by NT build 492. In order to eliminate influence of pore solution, a modification method of chloride diffusion coefficient was proposed. The modification method was developed based on the relationship between the electrical resistivity and chloride diffusion coefficient of concrete. It is reported that the modified chloride diffusion coefficient was much more consistent with the pore structure of specimens and closer correlated to the relative water absorption of concretes. This means that the effect of ionic concentration in pore solution is eliminated.

5.3.4.4 Breakthrough Time Method

Halamickova et al. (1995) first proposed to determine non-steady-state migration coefficient by measuring the breakthrough time. The migration coefficient is determined based on measurements of the time delay (breakthrough time), until the first significant increase in chloride concentration is detected in the downstream cell. The ratio c/c_0 is a small number where c is the first reliably detected chloride concentration in the downstream cell, as shown in Equation 5.21.

$$\frac{c}{c_0} = \frac{1}{2}\left[e^{ax}\text{erfc}\left(\frac{L + aD_{\text{nssm}}t_b}{2\sqrt{D_{\text{nssm}}t_b}}\right) + \text{erfc}\left(\frac{L - aD_{\text{nssm}}t_b}{2\sqrt{D_{\text{nssm}}t_b}}\right)\right] \tag{5.21}$$

where L is the thickness of specimen, and t_b is the breakthrough time, c is the first reliably detected downstream chloride concentration, c_0 is upstream chloride concentration. Through the same mathematical transformations, the above equation turns into,

$$D_{\text{nssm}} = \frac{RT}{zFE}\frac{L - \alpha\sqrt{L}}{t_b} \tag{5.22}$$

where

$$\alpha = 2\sqrt{\frac{RTL}{ZFU}} \times \text{erf}^{-1}\left(1 - \frac{2 \times c}{c_0}\right) \tag{5.23}$$

Halamickova et al. (1995) used the ratio $c/c_0 = 0.005$ in the calculation. There are also many other values that were used in the calculation. Yang et al. (2003)

used c/c_0 ratios of 0.001, 0.003, and 0.005; McGrath (1996) and Boddy et al. (2001) used a ratio of 0.003 as the breakthrough time for the calculation of chloride diffusion coefficient through the concrete specimens.

The drawbacks of this method include:

- Physical meaning of c/c_0 is not clear. Theoretically, c in Equation 5.21 refers to the chloride concentration at the surface of specimen at the downstream side. However, in practical calculation, c is assumed to be equal to the chloride concentration of the solution in downstream cell. Whether this assumption is valid is highly questioned.
- It takes quite long time for chloride ions to transport through concrete specimen, especially for low permeable concrete.
- There are many definitions of breakthrough time (c/c_0=0.001, 0.003, and 0.005). The definition of breakthrough time is illustrated in Figure 5.16. Different breakthrough time results in different results.

$$\text{When} \quad c/c_0 = 0.001, \alpha = 2\sqrt{\frac{RTL}{ZFU}} \times 2.185$$

$$\text{When} \quad c/c_0 = 0.003, \alpha = 2\sqrt{\frac{RTL}{ZFU}} \times 1.943$$

$$\text{When} \quad c/c_0 = 0.005, \alpha = 2\sqrt{\frac{RTL}{ZFU}} \times 1.821$$

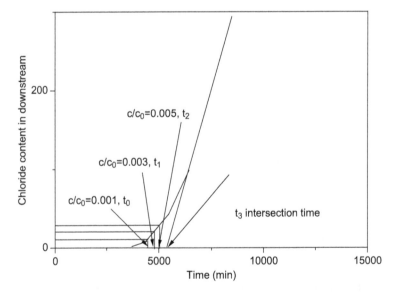

Figure 5.16 The definition of breakthrough time and intersection time.

5.3.4.5 Andrade and Castellote's Method

Andrade and Castellote carried out considerable research on chloride-related topics, and proposed several testing methods for the calculation of migration coefficient under both steady and non-steady conditions.

(1) Fitting Method

This method (Andrade 2000a) is based on fitting total chloride profiles to an analog diffusion equation to obtain D_{mig}:

$$c_t = c_{ts}.\mathrm{erfc}\frac{x}{2\sqrt{D_{mig}t}} \tag{5.24}$$

From D_{mig}, the chloride diffusion coefficient is calculated by means of the following expression:

$$D_{nssm} = \frac{RT}{zF}D_{mig}\frac{\ln L^2}{E} \tag{5.25}$$

where c is total chloride at certain depth, c_s is surface total chloride, L is the thickness of specimen, E is the applied voltage.

The drawbacks of this method are:

- This method requires laborious work for the determination of chloride concentrations at different layers.
- Wrong theory is used in this method. Diffusion theory is misused in this method.

Due to the drawbacks mentioned above, this method is seldom used. In the Round Robin test on testing methods for chloride transport in concrete initiated by Castellote and Andrade, this method is not included (Castellote 2006).

(2) Equivalent time method

Based on a rigorous solution of Nernst–Planck equation in steady-state conditions (Andrade 2000a), as shown in Equation 5.10, D_{nssm} can be obtained by Equations 5.26 to 5.28:

$$\frac{t}{t_{diff}} = \frac{6}{v^2}\left[v\coth\frac{v}{2} - 2\right] \tag{5.26}$$

$$D_{nssm} = \frac{x^2_d}{6t_{diff}} \tag{5.27}$$

$$v = \frac{zeE}{kT} \tag{5.28}$$

where e is the elementary charge, k is the Boltzman constant, x_d is the penetration depth measured by colorimetric method, and t_{diff} is the equivalent time for chloride ions to reach the same depth by diffusion alone.

Note that the thickness of specimens is not taken into consideration in Equations 5.26 to 5.28. In order to apply this method to specimens of any thickness, Castellote and Andrade carried out migration tests on specimens with different thicknesses, and correlated the thickness to a reference value of 10 cm. The migration results were also correlated to natural diffusion test. Then Equation 5.26 becomes (Castellote 2001b):

$$t_{diff} = \frac{t(v)\sqrt{\frac{10}{l}}}{\frac{6}{v^2}\left(v\coth\frac{v}{2} - 2\right)} \tag{5.29}$$

The main drawbacks of this method include:

- The equations are based on limited experimental results. This may not be applicable to a wide range of concrete.

(3) Conductivity method

Castellote and Andrade (2001a) proposed a method to calculate both non-steady-state and steady-state migration coefficients from one steady-state migration experiment through monitoring the conductivity of the downstream solution.

NT Build 355 needs expensive chemical analysis. Instead of the chemical analysis, Castellote and Andrade proposed to measure the chloride concentration in the anodic compartment by monitoring the conductivity of the solution in the anodic compartment. When it reaches steady state, the conductivity evolves with time linearly. A good relationship between the value of chloride concentration and conductivity was found:

$$\left[Cl^-\right](mmol/l) = -1.71 + 11.45 \times conductivity(25°C)(ms/cm) \tag{5.30}$$

By connecting a data logger to the conductivity electrode in the anodic compartment, the evolution of conductivity with time can be monitored continuously, and the migration coefficient can be calculated automatically by using Equation 5.13. The chemical analysis thus can be avoided.

The authors' results showed that a higher chloride concentration (higher than 0.5 mol/l) in the upstream compartment leads to a good consistence with natural diffusion tests, while a disagreement was found for lower chloride concentrations (lower than 0.5 mol/l). A twofold reasoning accounts for this (Castellote 2001a). On one hand, there are differences in the amount of chloride binding between migration and diffusion tests. At lower concentrations, the amount of chlorides bound in a migration test is lower than expected. However, at higher concentrations, combined chlorides are similar for migration and natural diffusion tests. On the other hand, below a concentration of 0.2 mol/l sodium chloride, the amount of chlorides is not enough to provide an optimum efficiency in the transport of the current by the chlorides. Therefore, a chloride solution with concentration higher than 0.5 mol/l is suggested in the test.

As mentioned above, D_{nssm} can also be determined from the same experiment through monitoring the conductivity of the downstream solution (Castellote 2001a). The conductivity of the solution is correlated with chloride concentration in the solution. The theoretical bases for obtaining D_{nssm} are the same as Equations 5.26–5.28. t_{int} is obtained by the intersection of the straight line of chloride flux characteristic of steady state with the x-axis. Thus, this method is called intersection time method. Instead of using a factor of 6 in Equation 5.28, a factor of 3 is applied. The results obtained are in good accordance with those of natural diffusion tests.

$$D_{nssm} = \frac{L^2}{3t_{int}} \qquad (5.31)$$

where L is the thickness of specimen, and t_{int} is the intersection time, as shown in Figure 5.15. It is an advantage that both non-steady and steady-state migration coefficients can be calculated from one experiment. However, there are still some concerns on this method:

- The relationship between conductivity of the solution in the downstream and chloride concentration may depend on the type of concrete. When chlorides are transported through concrete and into the downstream solution, they are accompanied by other ions. The amount and type of other ions depend on the type of concrete, which has a great influence on the conductivity of the solution;
- The factor of 3 in Equation 5.31 is based on limited experiments. This may have no universality.

5.3.4.6 Samson's Method

All the above mentioned electrically accelerated methods have some simplified assumptions in common:

- Constant electrical field assumption. That is to say, the electrical field along the concrete specimen follows a linear decay.
- Single species theory. Chloride ions transport in concrete independently, without interaction with other ions.

Samson et al. (2003) developed a quite complicated method for the determination of chloride migration coefficient which is based on the Nernst–Planck-Poisson equations. The interactions between various species and the distribution of electrical field along the specimen were taken into account by using the Nernst–Planck–Poisson equations.

When an electrical field is applied to the specimen with known initial and boundary conditions (concentrations and types of exposure solution, pore solution chemistry, porosity, etc.), the current passed through specimen can be calculated by coupling Nernst–Planck–Poisson equations and an assumed tortuosity. Ionic diffusion coefficient in concrete equals to the product of ionic diffusion coefficient at infinite dilution and the tortuosity. The error between the experimentally determined current and the model current can be calculated by using different tortuosities. The smallest error corresponding to the tortuosity leads to the best estimation of diffusion coefficient of the ions. This method needs to know the pore solution chemistry of concrete, which is not easy to get, especially for concrete with low water-to-cement ratio. The complexity of this method makes it not practical for engineering applications.

Samson's results showed that there are only slight differences between the electrical potential profiles predicted by the constant electrical field assumption and those predicted by the coupled Nernst–Planck–Poisson equations. This conclusion was also confirmed by Narsili (2007), and the differences between the two electrical potentials depend on the ionic concentration and type of the solution used in the test. However, slight differences in electrical field potentials lead to significant changes in chloride profiles (Samson 2003).

5.3.4.7 Friedmann's Method

Based on Nernst–Planck equation, electroneutrality, and measurements of current intensity, Friedmann et al. (2004) proposed an analytical model to calculate the effective migration coefficient.

The tests are carried out on a typical migration testing setup, and in two stages. At the first stage, both downstream and upstream cells are filled with NaOH (0.025 mol/l) and KOH (0.083 mol/l) solution, and measure the steady current without NaCl. The second stage consists of the measurement of steady electrical current after adding 0.5 mol/l NaCl in the cathodic solution, as shown in Figure 5.17. With the two steady current, and by considering electroneutrality, the migration coefficient can be computed as follows:

$$D_{ssm} = \frac{(i_i - i_f)RT}{c_0 E F^2 \left(\dfrac{D_{OH}}{D_{Cl}} - 1 \right)} \tag{5.32}$$

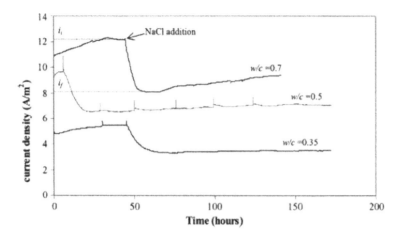

Figure 5.17 Chronoamperometry of the migration test for mortar. (From Fridemann 2004.)

where i_i is the steady current without NaCl, i_f is the steady current after adding NaCl. D_{OH} and D_{Cl} are hydroxyl and chloride diffusion coefficient at infinite dilution. c_0 is the chloride concentration in upstream cell.

The results of this method were compared with the results from NT build 355. It was found that the migration coefficient obtained with this method is one order of magnitude higher than that from NT build 355. The difference between the two values is due to the limits of NT build 355, because electroneutrality is obviously neglected in NT build 355 (Friedmann et al. 2004).

It is worth it to mention that Krabbenhøft et al. (2008) used coupled Nernst–Planck–Poisson equations to model the conventional migration tests, and found that the difference in the effective diffusivities computed on the basis of the single-species model, like NT build 492 and NT build 355, and the Nernst–Planck–Poisson equations typically amount to 50-100%. Krabbenhøft believed that by adjusting the ratio of NaCl to NaOH in cells, one can obtain a reasonable result by using the conventional simplified model.

The method has the following concerns:

- The judgment of steady state is not accurate by current measurement. Krabbenhøft et al. (2008) re-analyzed Friedmann's results by using Nernst–Planck–Poisson equations, and found that the steady state was not reached yet for the migration tests, while Friedmann considered them as steady state.
- Constant field assumption in the sample.
- The ratio of D_{OH}/D_{Cl} at infinite dilution may not equal to the ratio of D_{OH}/D_{Cl} at cement-based materials, and the former one is used in the calculation.
- It is developed only for steady-state tests.

5.3.5 Nernst–Einstein Equation

Nernst–Einstein is a particular case of Nernst–Planck equation (Bockris 1982). It has the form:

$$D = \frac{RT\sigma}{zF^2} \tag{5.33}$$

where D is the diffusion coefficient and σ is the conductivity. The most interesting practical aspect of this equation regarding concrete is the possibility of calculating effective diffusion coefficient from a single measurement of concrete conductivity or resistivity. Given that all the ions in the pore solution of concrete make contributions to the bulk conductivity of concrete, the part contributed by chloride ions can be expressed by transference number (Andrade 1993, Lu 1997):

$$t_{cl} = \frac{\sigma_{Cl}}{\sigma_b} \tag{5.34}$$

Thus, chloride effective migration coefficient can be calculated by,

$$D_{eff} = \frac{RT\sigma_{cl}}{zF^2} = \frac{RT\sigma}{zF^2} t_{cl} \tag{5.35}$$

where σ_{cl} is the partial conductivity of chloride, σ_b is the bulk conductivity of concrete. t_{cl} is the transference number of chloride.

The transference number of chloride is very difficult to quantify. Castellote et al. (1999b) studied the transference number of chloride for a migration test with NaOH+NaCl and distilled water in cathodic and anodic cells. In this case, chloride has different transference numbers at anodic and cathodic cells and specimen inside, because the chemistry composition is different in different parts. Thus, Castellote et al. (1999b) proposed different transference numbers, for example, anodic and cathodic transference numbers. The transference numbers also change with time due to the change in chemistry composition. To account for this, Castellote proposed differential transference number and accumulative transference number. However, all the transference numbers are not practical in the calculation of chloride migration coefficients.

From an engineering practical point of view, Andrade (2000) proposed a method to calculate the steady-state migration coefficient based on the measurement of resistivity:

$$D_{eff} = \frac{12 \times 10^{-5}}{\rho} \tag{5.36}$$

where ρ is the resistivity of concrete, and D_{eff} is the effective diffusion coefficient. Although this method is not directly from Nernst–Einstein equation, the relation between resistivity and migration coefficient is borrowed from Nernst–Einstein equation.

To overcome the variation of concrete pore solutions, Streicher and Alexander (1995) proposed a technique in which concrete was first dried and then saturated with a very concentrated (5 mol/l) NaCl solution. In this case, chloride and sodium ions are dominated in the pore solution, and the effect of other species on the conductivity can be ignored. The drawbacks of this technique will be discussed later. Based on Nernst–Einstein equation, Lu (1997, 1998, 1999) used the technique proposed by Streicher and Alexander in the experiment. Instead of transference number, Lu used different correction factors f to replace the transference number for different types of concrete in Equation 5.34. In addition to the drawbacks associated with the technique proposed by Streicher and Alexander, the main concern of this method is how to obtain the correction factor f for various types of concrete.

5.3.6 Formation Factor

The origin of formation factor is in geological research on saturated porous materials (Snyder 2001). For a nonconductive porous solid saturated with a conductive pore solution, the formation factor is the ratio of the pore solution conductivity to the bulk conductivity of porous materials (effect of solid microstructure and pore solution):

$$\Gamma = \frac{\sigma_p}{\sigma_b} \tag{5.37}$$

where τ is the formation factor, σ_p is the conductivity of pore solution, and σ_b is the bulk conductivity of porous materials.

Due to the relation between conductivity and diffusion coefficient as shown in Equation 5.33, formation factor is also used to estimate the chloride diffusion coefficient in concrete. The diffusivity of chloride ion through a saturated concrete can be correlated with the electrical conductivity of the solution and the saturated concrete as follows (Buenfeld and Newman 1987; Kyi and Batchelor 1994; Snyder 2001):

$$\Gamma = \frac{D_p}{D} = \frac{\sigma_p}{\sigma_b} \tag{5.38}$$

where D is the chloride diffusion coefficient in concrete, and D_p is the chloride diffusion coefficient in the pore solution.

Theoretically, it is a sound method to measure the diffusion coefficient of chloride ion in concrete. However, there are some concerns:

- Pore surface of concrete is negatively charged, not a non-conductive porous material.
- Pore solution of concrete is not easy to obtain.

- Pore solutions of concretes are concentrated solution, and vary in a wide range from 0.1 to 1 mol/l. Furthermore, pore solution changes with time due to the continuously hydration of cement, an exact value for D_p can be difficult to define.

As mentioned above, Streicher and Alexander (1995) proposed to first dry and then saturate concrete with 5 mol/l NaCl solution to avoid the difficulties in obtaining the pore solution and the chloride diffusion coefficient in the pore solution. They named the ratio of σ_p / σ_b as diffusibility ratio, which is actually formation factor. Then, the diffusion coefficient can be calculated by Equation 5.38.

The technique used to saturate concrete has some obvious concerns (Shi et al. 2007):

- Drying may lead to a change in microstructure.
- This technique also assumes that the concrete pore solution after saturation is the same as the NaCl solution used for saturating the concrete. This is not the case. The pore solution of concrete contains a wide range of ions (mainly alkali hydroxides), some of which will precipitate when the concrete is dried. When a solution enters into the concrete, the precipitated ions will then return to solution.
- The saturation of concrete with chloride solution may result in the formation of Friedel's salt and decrease the permeability of concrete.

Despite the drawbacks of formation factor method, Azad et al. (2019) believed that the formation factor can be used as an alternative to rapidly assess the transport properties of concrete; however, it does not consider chloride binding. Thus, they proposed a theoretical relationship to relate the formation factor, chloride binding parameters, and apparent chloride diffusion coefficient. Combining the formation factor with chloride binding properties provides an alternative to ASTM C1556-11a (2016) for determination of the apparent chloride diffusion coefficient and surface chloride concentration.

5.3.7 Other Methods

5.3.7.1 ASTM C 1202/AASHTO T227 Test Method

This test, originally developed by Whiting (1981), is often referred to as the Rapid Chloride Permeability Test (RCPT). In the RCPT test, a 50mm thick, 100mm in diameter concrete specimen which is saturated with limestone water is subjected to 60 V DC voltages. In the negative cell is a 3% NaCl and in the positive cell is a 0.3 mol/l NaOH solution. The chloride permeability of the concrete is rated based on the charge passed through the specimen during a six-hour testing period, using criteria in Table 5.2.

This method was adopted by the American Association of State Highway and Transportation Officials (AASHTO)—T277, Rapid Determination of the Chloride Permeability of Concrete—in 1983 and was designated by the American Society of Testing and Materials as ASTM C1202—Electrical Indication of Concrete's Ability to Resist Chloride Ion Penetration—in 1991. ASTM C1202 recognizes that a correlation between the rapid chloride permeability test method and the 90-day ponding test results is necessary, while AASHTO T277 does not require this correlation (Shi et al. 2007).

Many scientists and researchers all over the world have criticized this method during the past decade because of its lack of scientific bases and harsh testing conditions (Feldman et al. 1994; Pfeifer et al. 1994; Cao and Meck 1996; Scanlon and Sherman 1996; Shi et al. 1998; Shi 2004). The main criticisms are:

- Both chloride ions and other species are responsible for the charge passed through the specimen. The chemistry of pore solution has a remarkable effect on the conductivity of concrete but a minor effect on the permeability of concrete. Shi and his colleagues (Shi et al. 1998, Shi 2004) used electrochemical theory to quantitatively calculate how supplementary cementing materials affect the conductivity of pore solution of hardened cement pastes based on its chemistry. Also, the use of calcium nitrate as corrosion inhibitor may affect the conductivity of concrete significantly (Berke and Hicks 1992).
- The use of 60 voltages can cause a significant heating and increase the temperature of the test in most cases, which affect the test results greatly: The electrical conductivity of concrete increases with the increase of temperature. The significant increase of temperature can decompose some hydration products and cause cracking of the specimen, which then further increases the conductivity of the concrete. It is even worse for poor quality concrete due to the greater heating generated during the test.

In spite of its drawbacks, ASTM C1202 may still be useful for quality control to detect radical changes in the water-cementitious material-ratio (w/cm) or material properties. Some researches tried to improve ASTM C1202. McGrath (1996) found that the charge passed after 30 minutes in the RCPT test is as good an indicator of the permeability of the material as the six-hour charge passed for a total charge of less than 1,000 coulombs, and the 30-minute charge is a superior indicator of sample diffusion properties when total charge exceeds 1,000 coulombs since temperature is not built up. Feldeman et al. (1999) found that a very good relationship between the six-hour charge passed and the initial current or conductivity, and suggest to use the initial current or conductivity to replace the six-hour charge passed to evaluate the permeability of concrete. In this case, the test period is greatly shortened and the heating generated during the test may be avoided. Riding et al. (2008) proposed a simplified RCPT test, in which only the voltage drop across the sample was recorded, instead of the current. A good relationship was found between the simplified RCPT and RCPT.

In view of the fact that RCPT only provide the information about the conductivity of concrete, and has nothing to do with the concrete diffusion coefficient, Berke and Hicks (1992) established an empirical equation between charge passed and diffusion coefficient, as follows:

$$D = 0.0103 \times 10^{-8} Q^{0.84} \tag{5.39}$$

In order to overcome the drawback of RCPT which is sensitive to the chemical compositions of pore solution, Pilvar et al. (2015) proposed a new method called modified rapid chloride penetration test (MRCPT). In this new method, 23% NaCl solutions was used to saturate the specimen before test, and the uniformity of the pore solutions conductivity of the different concrete samples can be reached. Therefore, this method is less sensitive towards the variations in the pore solution conductivity in different concretes. The conductivity of the sample was used to evaluate the chloride resistance of concrete, instead of charge passed.

5.3.7.2 Salt Ponding Test (AASHTO T 259)

The ponding test—AASHTO T 259 (1980)—requires three slabs at least 75 mm thick and a surface area of 300 mm². These slabs are moist cured for 14 days, then conditioned in a room at 50% relative humidity for 28 days. After the conditioning, a 3% NaCl solution is ponded on the top surface, while the side surface is sealed and the bottom face is exposed to the drying environment at 50% relative humidity (as shown in Figure 5.18). After 90 days of ponding test, the chloride concentration of 0.5-inch thick slices from the slab is then determined (AASHTO T 259 1980). Typically, two or three slices are taken at progressive depths. The integral chloride contents to 41 mm after 90 days of ponding test are used for rating purposes.

People often measure chloride contents at several successive layers. The obtained profile can be used to fit error function solution to Fick's

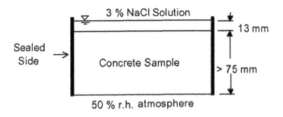

Figure 5.18 Illustration of AASHTO T 259 testing setup.

second law, and diffusion coefficient can be obtained (Berke and Hicks 1992; Andrade and Whiting 1996; Sherman et al. 1996; McGrath and Hooton 1999).

There are some obvious concerns about this method:

- The AASHTO T 259 90-day ponding test is intended to examine the combined influence of diffusion and capillary suction. However, the combination of these transport mechanisms may over- or underemphasize the importance of an individual process.
- In some applications, the chloride profile is measured at the increment of 0.5 inch (12.7mm). The increment is too large to provide a good description of the chloride profile. The use of this profile to fit error function solution to Fick's second may result in a great error in diffusion coefficient.

5.3.7.3 Water Pressure Method

Besides electrical field, another way to accelerate the transport of chloride is by applying pressure on one face that is in contact with chloride solution. Then, chloride transports by convection and diffusion. This will be governed by the following equation (Freeze and Cherry, 1979):

$$\frac{\partial c}{\partial t} = D \frac{\partial^2 c}{\partial t^2} - \bar{v} \frac{\partial c}{\partial x} \tag{5.40}$$

where $\bar{v}$ is the average linear rate of flow which is:

$$\bar{v} = -\frac{\phi}{\omega} \frac{\partial H}{\partial x} \tag{5.41}$$

where ϕ is the hydraulic permeability, ω is the porosity, and H is the applied pressure head. The solution to this differential equation is very similar to Tang's Equation 5.25:

$$\frac{c}{c_s} = \frac{1}{2} \left[\text{erfc} \left(\frac{x - \bar{v}t}{2\sqrt{Dt}} \right) + \exp \left(\frac{\bar{v}x}{D} \right) \text{erfc} \left(\frac{x + \bar{v}t}{2\sqrt{Dt}} \right) \right] \tag{5.42}$$

This allows the determination of chloride diffusion coefficients if a chloride profile is known at a specific time.

The testing setup is very similar to setup for water permeability, as shown in Figure 5.19. The side surface has to be sealed very well to avoid leakage. This method has a sound theoretical base, but it is seldom used. Because it needs special equipment, and high pressure may damage the microstructure of concrete.

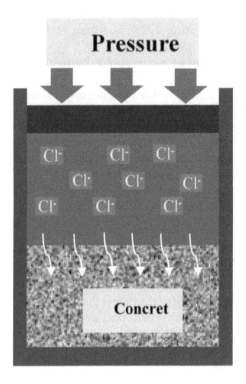

Figure 5.19 Schematic diagram of water pressure method.

5.3.7.4 AC Impendence Method

The AC impedance technique has long been used in the study of electro-chemical reactions. For a general electrochemical reaction by measuring the frequency response of the impedance Z, much information about the reaction mechanism and mass transfer can be obtained. It was also used in the determination of chloride diffusivity in concrete (Diaz 2006). It has some advantages over presumably simple and direct current measurements. For instance, in the case of two-probe measurements, impedance spectroscopy (IS) can separate voltage drops associated with the bulk sample and the electrodes. Ideally, if concrete behaves as a simple parallel RC circuit in series with the electrodes having its own impedance, then the total impedance is

$$Z_T = \frac{R_b}{1 + i\omega R_b C_b} + \frac{R_e}{1 + i\omega R_e C_e} \tag{5.43}$$

where R_b and C_b are the bulk resistance and capacitance, and R_e and C_e are the electrode resistance and capacitance. Figure 5.20 shows a plot of the real and imaginary impedance for the above circuit. The dip corresponds to

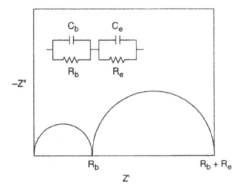

Figure 5.20 Impedance spectroscopy.

the resistance of the bulk, which is much lower than that of the electrodes. Such an approach can reduce the effect of the electrodes to more accurately determine the conductivity of concrete.

Again, with knowledge of the solute diffusivity and conductivity, the diffusivity of ions in cement-based materials may be determined using the Nernst–Einstein relation.

Shi (1999) used the coefficient of Warburg impedance to calculate the diffusion coefficient of chloride:

$$D = 3.54 \times 10^{-14} (A\sigma c)^{-2} \tag{5.43}$$

where A is the area of electrode surface, c is the chloride concentration, and σ is the coefficient of Warburg impedance.

Diaz (2006) proposed another method based on impedance spectroscopy to determine chloride diffusion coefficient in concrete. AC impedance method is a simple and fast method. However, compared to DC method, it attracted much less attention from researchers in the determination of chloride diffusivity over the past decades. This is probably because AC impedance method is simple and accurate to measure the resistance of concrete, but it is not easy to model the whole transport process. DC method is more straightforward and well understood.

5.4 STANDARDS ON TESTING METHODS FOR CHLORIDE TRANSPORT

Although many testing methods have been developed and proposed, just a few methods have been standardized. The testing methods adopted in standard or specifications in different countries are presented in this section, mainly the testing method standardized in America, Europe, and China.

(1) America

The AASHTO is probably the first to standardize the testing method for chloride transport: AASHTO T259 (1980), Standard Method of Test for Resistance of Concrete to Chloride Ion Penetration. This method later was adopted by America Society for Testing and Materials as ASTM C1543 in 2002: Standard Test Method for Determining the Penetration of Chloride Ion into Concrete by Ponding.

In 1983, AASHTO standardized another fast testing method which is first accelerated by electrical field: AASHTO T277, Standard Method of Test for Rapid Determination of the Chloride Permeability of Concrete. This method was also adopted by ASTM as ASTM C 1202 in 1994: Standard Test Method for Electrical Indication of Concrete's Ability to Resist Chloride Ion Penetration.

Following the European experiences, two other methods were adopted in American standards:

In 2003, AASHTO PT 64, Standard Method of Test for Predicting Chloride Penetration of Hydraulic Cement Concrete by the Rapid Migration Procedure; this method is similar to NT build 492

In 2004, ASTM C1556, Standard Test Method for Determining the Apparent Chloride Diffusion Coefficient of Cementitious Mixtures by Bulk Diffusion. This method is similar to NT build 443, and was upgraded to ASTM C1556-11a in 2016.

(2) Europe

Nordic countries have carried out extensive research works on chloride-related topics, and are pioneers in standardizing testing methods for chloride transport:

- NT BUILD 355 (1989), Concrete, Mortar and Cement Based Repair Materials: Chloride Diffusion Coefficient from Migration Cell Experiments, based on the principle of steady-state migration
- NT BUILD 443 (1995), Concrete, Hardened: Accelerated Chloride Penetration, based on the principle of non-steady-state diffusion under high chloride concentration
- NT BUILD 492 (1999), Concrete, Mortar and Cement-Based Repair Materials: Chloride Migration Coefficient from Non-Steady State Migration Experiments, based on the principle of non-steady-state migration

However, these methods are not European standards yet. Only one method was standardized as European standard:

- EN 13396 (since 2004), Products and Systems for the Protection and Repair of Concrete Structures—Test Methods—Measurement of

Chloride Ion Ingress. This method is similar to AASHTO T259 or ASTM C1543.

(3)   China

Two methods, i.e., NT build 492 and ASTM C 1202, are adopted in Chinese standard:

- GB/T 50082-2009, Standard for Test Methods of Long-Term Performance and Durability of Ordinary Concrete

5.5 RELATIONSHIP BETWEEN TEST RESULTS OBTAINED FROM DIFFERENT TEST METHODS

5.5.1 Non-Steady-State Migration and Diffusion Coefficients

There are five methods for the calculation of non-steady-state migration coefficient discussed in this chapter. They are NT build 492, breakthrough time method, equivalent time method (AC1), fitting method (AC2), and intersection time method (AC3). The latter three methods were proposed by Castellote and Andrade. The aim of migration test is to estimate the diffusion coefficient of concrete. It is necessary to compare these migration coefficients with diffusion coefficients. Since NT build 492 migration test method and NT build 443 diffusion test method are widely used, they were compared under the conditions specified in the standards. Table 5.3 gives the migration and diffusion coefficients obtained from NT build 492, NT build 443, AC1, and AC2 methods. It can be seen that the results from NT build 492 and AC1 were generally higher than the results from natural diffusion test. This is consistent with the results from other researchers (Castellote 2006).

The results from AC1 method have a very good linear relationship with the results from NT build 443, as shown in Figure 5.21. However, AC1 is based on rigorous solution of Nernst–Planck equation in steady-state

Table 5.3 Rating of Chloride Permeability of Concrete

Chloride Permeability	Charge (coulombs)
High	>4000
Moderate	2000-4000
Low	1000-2000
Very Low	100-1000
Negligible	<100

(ASTM C1202-2005).

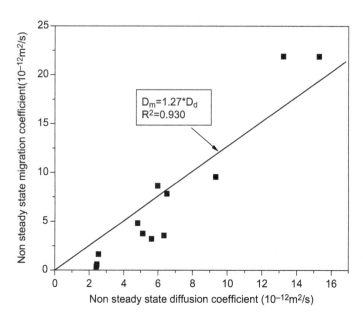

Figure 5.21 Relationship between migration coefficient obtained from ACI method and diffusion coefficient obtained from NT build 443. (From Yuan 2009.)

conditions. Using a solution obtained under steady-state condition in non-steady-state is theoretically inappropriate. What's more, AC1 method does not take the thickness of the specimen into account. This means AC1 method can be used only at the concentrations on which a relationship with diffusion test has been established. A different relationship might be obtained at other concentrations. This limits the application of AC1 method.

Furthermore, the thickness of the specimen is not considered, although it is needed to calculate the potential distribution in the specimen. This implies that the migration coefficient only is related to the potential drop through the specimen, but has nothing to do with the potential distribution in the specimen. To make up for this, Castellote and Andrade (2001b) proposed a reference thickness to address the potential distribution of specimens with different thicknesses based on a series of tests. The authors also emphasized that "the standardization equation presented here has been obtained directly from the experimental results presented here, and cannot be generalized until further experiments confirm its general validity."

The results from NT build 492 are generally 1.24 times higher than those obtained from NT build 443, as shown in Figure 5.22. Tang first proposed rapid chloride migration test in 1992. After several modifications and improvements on this method, it was standardized as NT Build 492 in 1999. This method has sounder theoretical basis than other methods, as it

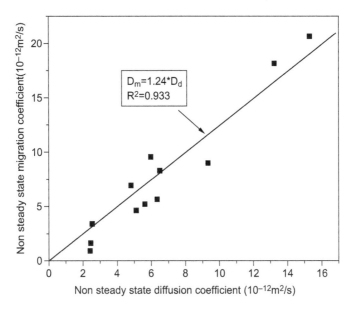

Figure 5.22 Relationship between migration coefficient obtained from NT build 492 and diffusion coefficient obtained from NT build 443. (From Yuan 2009.)

is based on basic electrochemical principles. However, the main drawbacks of this method include,

- It is not easy to measure the penetration depth by colorimetric method accurately by visual judgment. The color change boundary is very irregular, and it is not easy to distinguish it by naked eyes. Different operators may have different results.
- The chloride concentration at the color change boundary is affected by the alkalinity of concrete. The factor of 0.07 adopted in NT build 492 may not be suitable for all concrete.
- Single species theory. The influence of other ions on the migration is not taken into account.

It can be seen from Table 5.4 that the results from AC2 method, i.e., fitting method, were generally lower than that from NT build 443, and showed a much poorer linear relationship with diffusion test, see Figure 5.23. If looking at the theoretical bases behind the calculation equation, one can find that it is not strange to observe the discrepancy between AC2 fitting method and NT build 443. AC2 fitting method has the advantage of fitting closely the profiles obtained from specimen. This is, however, a misapplication of the underlying theories. The Nernst–Planck equation states that in an electrical potential field, ions will travel at some velocity to create a flux. This rate is controlled by migration coefficient. In the case of diffusion,

Table 5.4 Migration and Diffusion Coefficients Obtained from NT Build 492, NT Build 443, AC1 and AC2 Methods at 20°C($\times 10^{-12} m^2/s$)

| | Migration Tests | | | |
| Mix | NT Build 492 | AC1 Equivalent Time Method | AC2 Fitting Method | NT build 443 (Diffusion test) |
	10% (2 mol/l)	10% (2 mol/l)	10% (2 mol/l)	165 g/l (2.8 mol/l)
B6	18.15	21.89	9.03	13.24
B48	9.55	8.64	3	5.97
B35	5.65	3.57	1.08	6.35
FA6	20.64	21.88	–	15.3
FA48	8.99	9.57	–	9.34
FA35	3.39	1.63	1.45	2.55
SL6	6.92	4.83	2.12	4.81
SL48	8.29	7.83	1.45	6.5
SL35	1.62	0.59	0.61	2.46
SF6	5.21	3.23	2.46	5.62
SF48	4.63	3.77	1.4	5.12
SF35	0.91	0.36	–	2.42

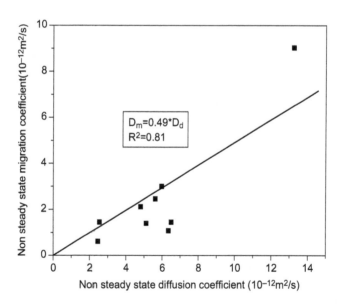

Figure 5.23 Relationship between migration coefficient obtained from AC2 fitting method and diffusion coefficient obtained from NT build 443. (From Yuan 2009.)

Table 5.5 Non-Steady-State Diffusion and Migration Coefficients Obtained from Different Methods at the Same Concentration ($\times 10^{-12} m^2/s$)

| | Non-Steady-State Migration Test (1 mol/l) | | | | | Diffusion (NT Build 443) |
| | Colorimetric Method | | Breakthrough Time Method | | | |
Mix	NT build 492	ACI	$c/c_0 = 0.003$	$c/c_0 = 0.005$	AC3	
B6	15.38	9.04	–	–	–	11.38
B48	10.66	9.7	5.76	5.46	11.77	8.31
B35	4.60	2.02	–	–	–	3.33
FA6	17.36	22.63	–	–	.	24.45
FA48	7.95	7.03	5.5	5.06	10.58	6.77
FA35	3	1.25	–	–	–	2.97
SL6	7.67	5.92	–	–	–	6.11
SL48	6.08	3.75	2.6	2.49	5.66	5.39
SL35	2.86	1.57	–	–	–	2.56
SF6	5.05	3.09	–	–	–	6.35
SF48	4.30	3.21	2.32	2.23	5	5.97
SF35	0.68	0.23	–	–	–	2.81

concentration gradients create a flux, and the rate of the flux is controlled by the diffusion coefficient. The migration coefficient cannot be substituted into the equations for diffusion, because migration is driven by electrical field, not concentration gradients.

Chloride migration coefficients are concentration dependent. However, the chloride concentrations specified in NT build 492 and NT build 443 are 10% (2 mol/l) and 165 g/l (2.8 mol/l), respectively. This concentration difference might lead to a misunderstanding of the relationship between these two methods. Table 5.5 gives non-steady-state diffusion and migration coefficients obtained from different methods at the same concentration (1 mol/l). At the same concentration level (1 mol/l), the relationship between the results from NT build 492 and the results from NT build 443 still follows a linear relationship. It can be seen that breakthrough time methods (both c/c_0=0.003 and 0.005) underestimate the diffusion coefficient. Yang et al. (2003) determined non-steady-state migration coefficient by defining breakthrough time as c/c_0=0.001, 0.003 and 0.005 respectively. The authors compared the migration coefficients with the diffusion coefficients determined from a 90-day ponding test, found that they were linearly related, and that migration coefficients were two times higher than diffusion coefficients determined from ponding test. It should be noticed that Yang et al. compared migration test with ponding test, instead of diffusion test.

Theoretically, the factor of c/c_0 is the ratio of chloride concentration at the upstream surface of the specimen to that at the downstream surface.

Breakthrough time method implicitly assumes that the chloride concentration at the upstream surface of the specimen is equal to the upstream concentration, and that the chloride concentration at the upstream surface equals the downstream concentration at breakthrough time. However, the actual chloride front is extremely irregular. When the downstream chloride concentration reaches $c/c_0=0.001$, 0.003 or 0.005, the amount of chloride passed through the specimen depends on the volume of downstream cell and the upstream chloride concentration. This is the reason why data from literature were quite different. Furthermore, it is also questioned whether the chloride concentration at the downstream surface of the specimen is equal to the downstream chloride concentration when the downstream chloride concentration reaches $c/c_0=0.001$, 0.003 or 0.005.

To address the relationship between non-steady-state migration and diffusion coefficients, Tang and Nilsson (1996) proposed an equation,

$$D_{nssd} = D_{nssm} \frac{\left(1 + K_b \dfrac{W_{gel}}{\omega}\right)}{1 + \dfrac{\partial c_b}{\partial c}} \tag{5.44}$$

where K_b is a binding constant (m³/kg$_{gel}$), W_{gel} is the amount of gel (kg$_{gel}$/m³$_{concrete}$) and ω is the capillary porosity. For K_b, values are given: for 100% Portland cement, 0.28×10^{-3} m³/kg$_{gel}$; for 30% slag and 70% Portland cement, 0.29×10^{-3} m³/kg$_{gel}$; and for 30% fly ash and 70% Portland cement, 0.32×10^{-3} m³/kg$_{gel}$. $\dfrac{\partial c_b}{\partial c}$ is a binding variable involved in immersion test, which is unknown. Based on Equation 5.44, Tang (1999) proposed a much more complicated theoretical equation to account for the relationship between these two coefficients. However, it is not practical, because many parameters in this equation are very difficult to experimentally determine.

Due to the wide application of NT build 443 and NT build 492, a simple linear relationship between non-steady-state migration coefficient determined from NT build 492 and non-steady-state diffusion coefficient determined from NT build 443 was proposed based on the experimental studies conducted on concretes with different water-to-binder ratios and different supplementary cementing materials in (Yuan 2009),

$$D_{nssm} = 1.2 \times D_{nssd} \tag{5.45}$$

5.5.2 Steady-State and Non-Steady-State Migration Coefficients

Yuan (2009) compared two methods, i.e., upstream method and downstream method (NT build 355), for steady-state migration coefficient (D_{ssm}).

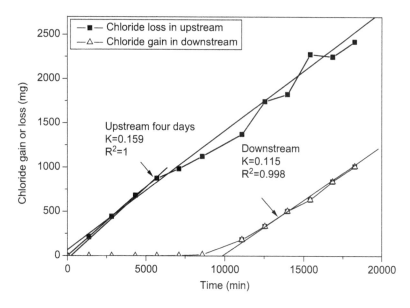

Figure 5.24 Evolution of chloride concentration in downstream and upstream with time. (From Yuan 2009.)

The typical evolution of downstream and upstream chloride concentrations with time is shown in Figure 5.24. During the steady-state migration test, the solutions in upstream and downstream were changed every four to five days to maintain a relatively constant upstream chloride concentration (not drop below 90% original concentration) and to avoid the neutralization in downstream solution. For the upstream method, ordinary concrete only needs two to three days' test duration to give a satisfactory result (Truc 2000). Therefore, in the upstream method, only the measurements of four days were used in the linear fit. Truc (2000) believed that the flux into concrete is constant from the beginning of test, and does not depend on chloride binding. However, when chlorides penetrate into concrete, part of them are in pore solution which contributes to the driving force, and part of them are bound to cement hydration products. It can be seen from Figure 5.24 that the slope of upstream method is greater than that of downstream method. This is because bound chloride is also counted as free chloride when using upstream method. It seems that the assumption that the flux into concrete is independent on chloride binding is invalid. This is further verified by the migration coefficients obtained from both methods, as shown in Table 5.6. It can be seen that the D_{ssm} obtained from the upstream method were generally greater than those from the downstream method.

It is observed that at the beginning of the test (for instance mix B48), the linear correlation of the data on the evolution of upstream chloride concentration was quite good, while at the latter stage (for instance, mix SL48), the

Table 5.6 Steady-State Migration Coefficients and Initial Current

Mix	Steady-State Migration ($\times 10^{-12} m^2/s$)				Initial Current (mA)	
	Upstream		Downstream (NT build 355)			
B48	0.641	0.67	0.464	0.5	37.6	39.4
	0.689		0.524		40.9	
	0.32(*)		0.517		39.6	
FA48	0.684	0.76	0.431	0.45	31.2	31.6
	0.811		0.464		31.8	
	0.778		0.46		31.8	
SL48	0.525	0.46	0.189	0.21	23	22.9
	0.387		0.225		22.9	
SF48	0.410	0.29	0.282	0.25	12.2	11.2
	0.161		0.220		10.1	

* This value is not adopted.

Table 5.7 Steady-State Migration and Non-Steady-State Migration/Diffusion Coefficients ($\times 10^{-12} m^2/s$)

Mix	Steady-State Migration		Non-Steady-State Migration	Non-Steady-State Diffusion
	Upstream Method	NT Build 355		
B48	0.65	0.5	10.66	8.31
FA48	0.74	0.41	7.95	6.77
SL48	0.46	0.21	6.08	5.39
SF48	0.29	0.22	4.30	5.97

data varied quite significantly. This is not well understood. The upstream chloride concentration (1 mol/l) was much higher compared to the titration solution (0.01 mol/l). Thus, the solution pipetted from upstream cell for the determination of the concentration of the solution had to be diluted. This affects the precision of upstream method. In contrast, the data on the evolution of downstream chloride concentration always showed a good linear relationship.

Compared to the non-steady-state migration coefficients, steady-state migration coefficients are one order of magnitude lower, as shown in Table 5.7. Tang (2001) also observed similar phenomena. Tang and Nilsson (1996) proposed an equation to account for this:

$$D_{ssm} = D_{nssm}\left(\omega + k_b \cdot W_{gel}\right) \tag{5.46}$$

For the calculation of W_{gel}, the following equation is used (Audenaert 2007):

$$W_{gel} = 1.25\,h\,C \tag{5.47}$$

where h is the degree of hydration and C is the cement content (kg/m³). The test specimens are stored until the testing age in a climate room at $20 \pm 2°C$ and at least 90% R.H. for at least 28 days. This means that the degree of hydration will not strongly differ from the ultimate degree of hydration, which could be determined by the Mill formula (Audenaert 2007):

$$h_{ultim} = \frac{1.031\,\dfrac{W}{C}}{0.194 + \dfrac{W}{C}} \tag{5.48}$$

Take mix B48, for example; insert all the values to Equation 5.46: K_b is 0.28×10^{-3} m³/kg$_{gel}$, ε is 14.1%, the calculated steady-state migration coefficient is 2.54×10^{-12} m²/s, and the measured value was 0.5×10^{-12} m²/s. The theoretical value is five times the experimental value.

Tang and Nilsson (1999, 2001) proposed a more complicated model to describe the relationship between D_{ssm} and D_{nssm},

$$D_{nssm} = \frac{D_{ssm} + \left(\dfrac{1}{a}\dfrac{\partial c}{\partial x} + c\right)\dfrac{\partial D_{ssm}}{\partial c}}{\omega\left[1 + \left(\dfrac{\partial c_b}{\partial c}\right)_m\right]} \tag{5.49}$$

where $\dfrac{1}{a}\dfrac{\partial c}{\partial x} = -0.06$ mol / l, $c = 0.1$ mol / l, $\left(\dfrac{\partial c_b}{\partial c}\right)_m = k_{bm}\dfrac{W_{gel}}{\varepsilon}$, $k_{bm} = 0.59 \times$

10^{-3}, W_{gel} is the weight of gel, ω is the porosity by volume, $\dfrac{\partial D_{ssm}}{\partial c} = -D_{ssm}\dfrac{B_m}{B_m + 55.46}$, $B_m = f\left(1 + |\beta_v|\right) - 1$. $|\beta_v| = 0.244$.

According to Tang and Nilsson (1999, 2001a), f is the core parameter that reflects the transport property of concrete, which varies in different concretes. It can be assumed that $f = 11,000$, $W_{gel} = 350$ kg/m³, and $\varepsilon = 0.141$. Inserting these values into Equation 5.49, $D_{ssm} = 0.34 \times D_{nssm}$ is obtained. However, it does not fit the experimental results, as shown in Table 5.8. This might be because the core parameter f is not the right one. According to Tang, f is obtained by measuring migration coefficients at different concentrations.

Spiesz et al. (2013) found that the D_{nssm} is identical to the intrinsic chloride migration coefficient in the pore solution (D_0) and with the effective chloride migration coefficient (D_{ssm}) divided by the porosity (D_{ssm}/φ).

Table 5.8 Comparison between D_{nssm} and D_{ssm}($\times 10^{-12}$m^2/s) (Tang)

w/c	0.35	0.40	0.50	0.75
D_{nssm}	1.1	1.6	2.1	3.2
D_{ssm}	2.8	5.4	13.6	39.7

5.5.3 ASTM C1201 (or Initial Current) and Migration Coefficients

In NT build 492, the initial current with the voltage pre-set at 30 V is measured for each specimen. According to the measured current, the testing voltage is selected. Initial current is also an indicator of chloride resistance of concrete (Feldman 1999). In Yuan (2009), NT build 492 test was carried out on a broad range of concrete under different conditions at the age of 56 days. Some mixes were also tested at 180 days. Since the initial current is related to the testing temperature, and upstream solution concentration, the initial current can only be compared at the same temperature and upstream solution. The relationship between initial current and migration coefficient was studied on the concretes subjected to the conditions of 10% NaCl upstream solution and 20°C.

Figure 5.25 gives the relationship between initial current with the voltage of 30V and migration coefficient. The results of some mixes at 180 days are also given. A good linear relationship can be found, as shown in Figure 5.25.

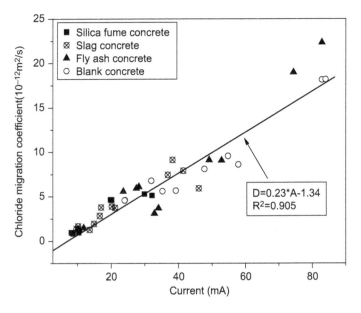

Figure 5.25 Relationship between initial current and Dnssm. (From Yuan 2009.)

It is well known that the initial current of concrete at a given voltage is related with the resistivity or conductivity of concrete. An empirical relationship between resistivity of concrete and diffusion coefficient was given by Berke and Hicks (1992),

$$D = 54.6 \times 10^{-8} \rho^{-1.01} \tag{5.50}$$

where ρ is the resistivity of concrete. It can be seen that the diffusion coefficient and resistivity can also be approximately expressed in linear relationship.

Andrade and Whiting (1996) used the initial resistivity in the calculation of diffusion coefficient by using Nernst–Einstein equation, and found that the concrete rankings obtained from this method are similar to those obtained from 90-day ponding tests.

Although ASTM C 1202 is criticized by many researchers, as reviewed above, it is still widely used in practice. Initial current was also linearly related to the charge passed through specimens within six hours (Feldman 1999), which is the ranking parameter in ASTM C 1202.

Although a good linear relationship was obtained between initial current and migration coefficient, according to Shi (2004) the electrical conductivity or resistivity of concrete cannot be used as an indicator of their chloride permeability. Because the initial current (or resistivity) of concrete depends on two parameters, the pore solution and the pore structure of concrete, while the permeability of concrete depends on the pore structure, the chemistry of pore solution has little to do with the transport of chloride but has great effects on electrical conductivity of concrete. Thus, when ranking the chloride resistance of concrete with initial current, special cares should be paid to the concretes with completely different chemistry of pore solutions, especially in the case of low w/b ratio.

5.5.4 NT Build 443 Results Obtained from Free and Total Chlorides

NT build 443 is widely used. The error function solution to Fick's second law is also widely used for engineering applications, and can be rewritten as,

$$\frac{c_s - c}{c_s - c_i} = \mathrm{erf}\left(x / \sqrt{4D_{\mathrm{app}}t}\right) \tag{5.51}$$

when $c_i = 0$, Equation 5.41 turns into,

$$\frac{c}{c_s} = 1 - \mathrm{erf}\left(x / \sqrt{4D_{\mathrm{app}}t}\right) \tag{5.52}$$

According to Fick's second law, the concentrations in Equation 5.52 refer to free chlorides. Total chloride can replace free chloride only if there is no binding or free chloride is linear with total chloride and the line has to pass through the origin (Tang and Nilsson 1996). As stated above, free and bound chlorides follow a non-linear relationship. This means,

$$\frac{c_t}{c_{st}} = \frac{c_f + \alpha c^{\beta}}{c_{sf} + \alpha c_s^{\beta}} \neq \frac{c_f}{c_{sf}} \tag{5.53}$$

where the subscripts f and t correspond to free and total chlorides; subscripts st and sf represent surface total and surface free chlorides. Accordingly, total chloride cannot be used to replace free chloride. In NT build 443 and engineering practices, however, total chloride profile is used. Therefore, there exists a theoretical error in NT build 443. To evaluate the error caused by using total chloride, both free and total chloride profiles were used to fit the error function solution to Fick's second law. Table 5.9 gives the diffusion coefficients obtained from fitting both total and free chloride profiles of concrete at different conditions. A very good linear relationship with a slope of 0.76 between diffusion coefficients obtained from fitting total and free chloride profiles was reached, as shown in Figure 5.26.

Table 5.9 Diffusion Coefficients Obtained from Fitting Total and Free Chloride Profiles of Concrete at Different Conditions ($\times 10^{-12} m^2/s$)

Mix	165 g/l Nacl, 5°C		165 g/l Nacl, 20°C		165 g/l Nacl, 40°C		1 mol/l Nacl, 20°C	
	Total	Free	Total	Free	Total	Free	Total	Free
B6	8.12	6.49	13.24	10.52	55.3	40.62	11.38	12.86
B48	5.38	4.4	5.97	4.95	12.8	10.34	8.31	7.95
B35	2.31	2.74	6.35	5.07	7.22	4.12	3.33	2.97
FA6	25.2	17.25	15.3	9.49	10.6	7.66	24.45	17.6
FA48	4.8	4.36	9.34	6.54	7.93	5.71	6.77	6.43
FA35	2.47	1.86	2.55	2.25	3.73	2.91	2.97	2.75
SL6	4.16	4.06	4.81	4.68	10.4	8.29	6.11	5.05
SL48	6.56	6.61	6.5	5.77	5.44	4.32	5.39	3.38
SL35	3.3	2.7	2.46	2.32	4.76	4.08	2.56	2.14
SF6	6.22	4.93	5.62	4.91	8.55	7.77	6.35	6.58
SF48	3.77	2.71	5.12	3.84	1.66	1.98	5.97	5.1
SF35	1.6	1.34	2.42	2.33	4.27	4.35	2.81	3.38

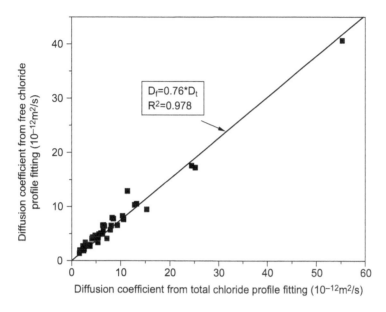

$D_f=0.76*D_t$
$R^2=0.978$

Figure 5.26 Relationship between diffusion coefficients obtained from fitting total and free chloride profiles. (From Yuan 2009.)

5.6 SUMMARY

Many testing methods have been developed for the transport of chloride in concrete. However, each method has its disadvantages and advantages. To be a good test method, it should fulfill the following requirements:

- Clear and right theoretical base
- Simple operations
- Reproduce reality
- Fast results
- Good reproducibility and repeatability

However, it seems no one method can meet all the requirements. So far, NT build 492 seems the most suitable test method for the non-steady-state migration test. The major theoretical drawback of many electrically accelerated methods is the affect of other species on the migration process is not taken into account, i.e., single species theory. The application of coupled Nernst–Planck and Poisson equations makes the theory more scientific sound, but it is too complicated to be used in engineering applications. It seems that Friedmann's analytical solution for Nernst–Planck equation coupled with electroneutrality is quite promising. Efforts still have to be devoted to make the experiment more practical for engineers.

Fitting the total chloride profile to the error function solution to Fick's second law is theoretically oversimplified. Free chloride should be used. The non-steady-state diffusion coefficient obtained by curve fitting of total chloride was linearly related with that of free chloride by a factor of 0.76.

Of the two steady-state migration tests, the results from downstream method (i.e., NT build 355) were generally lower than that from upstream method. This was because bound chloride was also counted as free chloride when using upstream method. Furthermore, the precision of upstream method is poorer than downstream method.

Of the five non-steady-state migration test methods, the results from NT build 492 and AC1 method were well linearly related to that from NT build 443: $D_{nssm} = \lambda D_{nssd}$, the factor λ was dependent on w/b ratio and supplementary cementing materials; AC2 method (i.e., fitting method) gave the worst estimation of diffusion coefficient; the results of AC3 method were slightly higher than that from NT build 443; the results of breakthrough method was generally lower than that from diffusion test. The chloride concentrations specified in NT build 492 (10%) and NT build 443(165 g/l) are different. When the two tests were carried out at the same concentration of 1 mol/l, the linear relationship still held.

Compared to non-steady-state migration coefficients, steady-state migration coefficients were always around one order of magnitude lower.

REFERENCES

AASHTO T 277-86, Rapid Determination of the Chloride Permeability of Concrete. American Association of States Highway and Transportation Officials. Standard Specifications—Part II Tests, Washington, DC; 1990.

AEC Laboratory. Concrete Testing, Hardened Concrete. Chloride Penetration. APM 302, 2nd ed. AEC Laboratory, Vedbek, Denmark; May 1991.

Andrade C, Whiting D. A comparison of chloride ion diffusion coefficients derived from concentration gradients and non-steady state accelerated ionic migration. *Materials and Structures.* 1996, 29:476–484.

ASTM C1202. Electrical indication of concrete's ability to resist chloride ion penetration. Annual Book of American Society for Testing Materials Standards; 2000, 04:02.

ASTM C1556-11a, Standard test method for determining the apparent chloride diffusion coefficient of cementitious mixtures by bulk diffusion, ASTM, West Conshohocken, PA; 2016.

Andrade C. Calculation of chloride diffusion coefficients in concrete from ionic migration measurements. *Cement and Concrete Research.* 1993, 23:724–742.

Andrade C, Castellote M, Alonso C, González C. Non-steady-state chloride diffusion coefficients obtained from migration and natural diffusion tests. Part 1: Comparison between several methods of calculation. *Materials and Structures.* 2000, 33:21–28.

Andrade C, Sanjuan MA. Experimental Procedure for the calculation of chloride diffusion coefficients in concrete from migration tests. *Advances in Cement Research*. 1994, 6:127–134.

Arsenault J, Gigas JP, Ollivier JP. Determination of chloride diffusion coefficient using two different steady-state methods: Influence of concentration gradient. Proceedings of the International RILEM Workshop, Nilsson LO and Ollivier JP (eds.); 1995, 150–160.

Audenaeret K, Boel V, De Schutte G. Chloride migration in self compacting concrete, Proceedings of the fifth international conference on concrete under severe conditions, CONSEC'07, , Tours, France, 4–6 June 2007, 291–298.

Azad VJ, Erbektas AR, Qiao CY. Relating the formation factor and chloride binding parameters to the apparent chloride diffusion coefficient of concrete. *Journal of Materials in Civil Engineering*. 2019, 31:04018392.

Berke NS, Hicks MC. Estimating the life cycle of reinforced concrete decks and marine piles using laboratory diffusion and corrosion data in corrosion forms and control for infrastructure. *ASTM STP* 1139, 1992:207–231.

Bockris JM, Conway BE. *Modern Aspects of Electrochemistry*, 14th edition. Plenum, New York; 1982.

Boddy A, Hooton RD, Gruber KA. Long-term testing of the chloride-penetration resistance of concrete containing metakaolin. *Cement and Concrete Research*. 2001, 31:759–765.

Buenfeld NR, Newman JB. Examination of three methods for studying ion diffusion in cement pastes, mortars and concrete. *Materials and Structure*. 1987, 20:3–10.

Byfors K. Influence of silica fume and flyash on chloride diffusion and pH values in cement pastes. *Cement and Concrete Research*. 1987, 17:115–130.

Cao HT, Meck E. A review of the ASTM C1202 standard test. *Concrete In Australia*. 1996:23–26.

Castellote M, Andrade C. Round-Robin test on methods for determining chloride transport parameters in concrete. *Materials and Structures*. 2006, 39:955–990.

Castellote M, Andrade C, Alonso C. Chloride binding isotherms in concrete submitted to non-steady-state migration experiments. *Cement and Concrete Research*. 1999a, 29:1799–1806.

Castellote M, Andrade C, Alonso C. Modelling of the processes during steady-state migration tests: Quantification if transference numbers. *Materials and Structures*. 1999b, 32:180–186

Castellote M, Andrade C, Alonso C. Measurement of the steady and non-steady-state chloride diffusion coefficients in a migration test by means of monitoring the conductivity in the anolyte chamber Comparison with natural diffusion tests. *Cement and Concrete Research*. 2001a, 31:1411–1420.

Castellote M, Andrade C, Alonso C. Non-steady-state chloride diffusion coefficients obtained from migration and natural diffusion tests. Part II: Different experimental conditions joint relations. *Materials and Structures*. 2001b, 34:323–331.

Crank J. *The Mathematics of Diffusion*, 2nd ed. Oxford University Press, London; 1975.

Chatterji S. On the non-applicability of unmodified Fick's laws to ion transport through cement based materials. In: Nilsson LO, Ollivier JP (eds.). *Proceedings of the International RILEM Workshop-Chloride Penetration Into Concrete*, Paris; 1995, 64–73.

Díaz B, XR No´voa, MC Pe´rez. Study of the chloride diffusion in mortar: A new method of determining diffusion coefficients based on impedance measurements. *Cement and Concrete Composites* 2006, 28:237–245.

Feldman RF, Chan GW, Brousseau RJ, Tumidajski PJ. Investigation of the rapid chloride permeability test. *ACI Materials Journal.* 1994, 91:246–255.

Feldman RF, Luiz R, Prudencio JU, Chan G. Rapid chloride permeability test on blended cement and other concretes: Correlations between charge, initial current and conductivity. *Construction Building Materials.* 1999, 13:149–154.

Friedmann H, Amiri O, Aı¨t-Mokhtar A, Dumargue P. A direct method for determining chloride diffusion coefficient by using migration test. *Cement and Concrete Research.* 2004, 34:1967–1973.

Freeze RA, Cherry JA. *Groundwater.* Prentice-Hall, New Jersey; 1979.

Gao PH, Wei JX; Zhang TS, Hu J, Yu Q. Modification of chloride diffusion coefficient of concrete based on the electrical conductivity of pore solution. *Construction and Building Materials,* 2017, V145:361–366.

Glass GK, Wang Y, Buenfeld NR. An investigation of experimental methods used to determine free and total chloride contents. *Cement and Concrete Research.* 1996, 26(9):1443–1449.

Halamickova P, Detwiler RJ, Bentz DP, Garboczi EJ. Water permeability and chloride ion diffusion in portland cement mortars: Relationship to sand content and critical pore diameter. *Cement and Concrete Research.* 1995, 25:790–802.

Hanson CM, Sorensen B. The influence of cement fineness on chloride diffusion and chloride binding in hardened cement paste. *Nordic Concrete Research.* 1987:57–72.

Haque MN, Kayyali OA. Free and water soluble chloride in concrete. *Cement and Concrete Research.* 1995, 25:531–542.

Krabbenhøft K, Krabbenhøft J. Application of the Poisson-Nernst-Planck equations to the migration test. *Cement and Concrete Research.* 2008, 38:77–88.

Lu X. Application of the Nernst-Einstein equation to concrete. *Cement end Concrcte Research.* 1997, 27(2):293–302.

Lu X. Rapid determination of the chloride diffusivity in concrete. Proceedings of the Second International Conference on Concrete under Severe Conditions (CONSEC'98), Tromso, Norway, 21–24 June 1998, 1963–1969.

Lu X. A rapid method for evaluation of the permeability of concrete. Chinese Patent CN1249427A; 1999.

McGrath P. Development of test methods for predicting chloride penetration into high performance concrete. PhD Thesis, Department of Civil Engineering, University of Toronto; 1996.

McGrath P, Hooton RD. Influence of voltage on chloride diffusion coefficients from chloride migration tests. *Cement and Concrete Research.* 1996, 26:1239–1244.

McGrath P, Hooton RD. Re-evaluation of the AASHTO T259 90-day salt ponding test. *Cement and Concrete Research.* 1999, 29:1239–1248.

Narsili GA, Li R, Pivonka P, Smith DW. Comparative study of methods used to estimate ionic diffusion coefficients using migration tests. *Cement and Concrete Research.* 2007, 37:1152–1163.

Nilsson LO. Concepts in chloride ingress modelling, Third RILEM Workshop on Testing and Modelling the Chloride Ingress Into Concrete, Madrid, Spain, 9–10 September 2002, 29–48.

Otsuki N, Nagataki S and Nakashita K. Evaluation of AgNO3 solution spray method for measurement of chloride penetration into hardened cementitious matrix materials. *ACI Materials Journal.* 1992, 89(6):587–592.

Page CL, Short NR, El-Tarras A. Diffusion of chloride ions in hardened cement pastes. *Cement and Concrete Research.* 1981, 11:395–406.

Park B, Jang SY, Cho J-Y, Kim JY. A novel short-term immersion test to determine the chloride ion diffusion coefficient of cementitious materials. *Construction and Building Materials.* 2014, 57:169–178.

Pfeifer D, McDonald D, Krauss P. The rapid chloride test and its correlation to the 90-day chloride ponding test. *PCI Journal.* 1994:38–47.

Pilvar A, Ramezanianpour AA, Rajaie H. New method development for evaluation concrete chloride ion permeability. *Construction and Building Materials.* 2015, 93:790–797.

Riding KA, Poole JL, Schindler AK, Juenger MCG, Folliard KJ. Simplified concrete resistivity and rapid chloride permeability test method. *ACI Materials Journal.* 2008, 105(4):390–394.

Samson E, Marchand J, Snyder KA. Calculation of ionic diffusion coefficients on the basis of migration test results. *Materials and Structures.* 2003, 36:156–165.

Scanlon JM, Sherman MR. Fly ash concrete: An evaluation of chloride penetration test methods. *Concrete International.* 1996, 18:57–62,.

Sherman MR, McDonald D, Pfeifer D. Durability aspect of precast prestressed concrete. Part 2: Chloride permeability study. *PCI Journal.* 1996:76–95.

Shi C. Formation and stability of 3CaO.CaCl2.12H2O. *Cement and Concrete Research.* 2001, 31:1273–1275.

Shi C. Effect of mixing proportions of concrete on its electrical conductivity and the rapid chloride permeability test (ASTM C1202 or ASSHTO T277) results. *Cement and Concrete Research.* 2004, 34:537–545.

Shi C, He F. A thin layer of concrete sampling machine. Chinese Patent CN10629874; 2010.

Shi C, Stegemann JA, Caldwell R. Effect of supplementary cementing materials on the rapid chloride permeability test (AASHTO T 277 and ASTM C1202) results. *ACI Materials.* 1998, 95:389–394.

Shi C, Yuan Q, Deng D, Zheng K. Test, methods for the transport of chloride ion in concrete. *Journal of the Chinese Ceramic Society.* 2007, 35:522–530.

Shi M, Chen Z, Sun J. Determination of chloride diffusivity in concrete by AC impedance spectroscopy. *Cement and Concrete Research.* 1999, 29:1111–1115.

Snyder KA. The relationship between the formation factor and the diffusion coefficient of porous materials saturated with concentrated electrolytes: Theoretical and experimental considerations. *Concrete Science and Engineering.* 2001, 3:216–224.

Spiesz P, Brouwers HJH. The apparent and effective chloride migration coefficients obtained in migration tests. *Cement and Concrete Research.* 2013, 48:116–127.

Stanish K, Hootonb RD, Thomas MDA. Testing the chloride penetration resistance of concrete: A literature review, No. FHWA Contract DTFH61-97-R-00022, United States. Federal Highway Administration; 2001.

Stanish K, Hootonb RD, Thomas MDA. A novel method for describing chloride ion transport due to an electrical gradient in concrete: Part 1. Theoretical description. *Cement and Concrete Research.* 2004, 34:43–49.

Streicher PE, Alexander MG. A critical evaluation of chloride diffusion test methods for concrete. Third CANMET/ACI International Conference on Durability of Concrete, Supplementary Papers, Nice, France, 1994, 517–530.

Streicher PE, Alexander MG. A chloride conduction test for concrete. *Cement and Concrete Research*. 1995, 25:1284–1294.

Tang L, Nilsson L-O. Chloride diffusivity in high strength concrete. *Nordic Concrete Research*. 1992b, 11:162–170.

Tang L, Nilsson L-O. Service life prediction for concrete structures under sea water by a numerical approach. Proceedings of the 7th International Conference on Durability of Building Materials and Components, Stochholm, Sweden, May 19–23, 1996, 1:97–106.

Tang L. Concentration dependence of diffusion and migration of chloride ions. Part 1. Theoretical considerations. *Cement and Concrete Research*. 1999; 29:1463–1468.

Tang L, Sørensen HE. Precision of the nodic test methods for measuring the chloride diffusion/migration coefficients of concrete. *Materials and Structures*. 2001, 34:479–485.

Tetsuya I, Shigeyoshi M, Tsuyoshi M. Chloride binding capacity of mortars made with various Portland cements and mineral admixtures. *Journal of Advanced Concrete Technology*. 2008, 6(2):287–301.

Truc O, Ollivier JP, Carcassès M. A new way for determining the chloride diffusion coefficient in concrete from steady state migration test. *Cement and Concrete Research*. 2000, 30:217–226.

Vladimir P. Water extraction of chloride, hydroxide and other ions from hardened cement pastes. *Cement and Concrete Research*. 2000, 30:895–906.

Voinitchi D, Julien S, Lorente S. The relation between electrokinetics and chloride transport through cement-based materials. *Cement and Concrete Composites*. 2008, 30:157–166.

Whiting D. Rapid determination of the chloride permeability of concrete. Research Report FHWA/RD-81/119; 1981.

Yang CC, Cho SW. An electrochemical method for accelerated chloride migration test of diffusion coefficient in cement-based materials. *Materials Chemistry and Physics*. 2003, 81:116–125.

Yuan Q. Fundamental studies on test methods for the transport of chloride ions in cementitious materials. PhD thesis. Ghent University, Belgium; 2009.

Zhang T, Gjorv OE. An electrochemical method for accelerated testing of chloride diffusion. *Cement and Concrete Research*. 1994, 24:1534–1548.

Zhang T, Gjrøv OE. Effect of ionic interaction in migration testing of chloride diffusivity in concrete. *Cement and Concrete Research*. 1995, 25(7):1535–1542.

Chapter 6

Determination of Chloride Penetration in Cement-Based Materials Using AgNO₃-Based Colorimetric Methods

6.1 INTRODUCTION

Chloride-induced steel corrosion is the main reason for deterioration of reinforced concrete structures exposed to marine environments and/or de-icing salts. When a critical chloride concentration is reached near the surface of steel reinforcement, corrosion of the steel may occur as a result of the local breakdown of its passivated layer formed at high alkaline environment of concrete pore solution. The corrosion of reinforced concrete structures is normally described as a two-stage process: (i) corrosion initiation stage, referring to the time that a critical chloride concentration is reached at the surface of steel reinforcement; and (ii) corrosion propagation stage, starting from corrosion until the damage of concrete structures is beyond the acceptable limits. Considering the fast development of the corrosion propagation stage and for the sake of safety, the time required for corrosion initiation stage is usually used for design and prediction of the service life of reinforced concrete structures. The progression of chloride penetration front over exposure time under a certain chloride environment can be used to monitor the ingress of chloride ions into concrete and for service life prediction (Stanish 2002). An accurate method to determine the chloride penetration front is to acquire the total chloride profile, represented by the total chloride content as a function of the chloride ingress depth. However, this method is very time consuming as it requires significant working load, such as profile grinding and chemical analysis of the ground samples from different depths. A much easier and less costly approach to obtain the chloride penetration front is to use AgNO₃-based colorimetric methods, which requires direct spay of AgNO₃-based solution on the freshly split concrete surface. The color change boundary after spray of AgNO₃-based solution can be used as an indication of the chloride penetration front. This method is often used for rapid measurement of chloride ion penetration depth in concrete exposed to chloride environments. In this chapter, the procedure and applicability of the three different AgNO₃ colorimetric methods are described and discussed.

Table 6.1   Use of AgNO$_3$+Fluorescein and AgNO$_3$+ K$_2$CrO$_4$ in Silver Measuring
Methods

Method	AgNO$_3$ + K$_2$CrO$_4$	AgNO$_3$ + Fluorescein
Indicator	K$_2$CrO$_4$	Fluorescein
Titration solution	AgNO$_3$ solution	AgNO$_3$ solution
pH value of the solution	6.5–10.5	7–8
Color change at titration end-point	White→brick red	Yellowish green →pink
Reference	Mohr (1856)	Fajans (1924)

Li (2005)

6.2 DETERMINATION OF CHLORIDE INGRESS DEPTHS

Since the 1970s, three AgNO$_3$ colorimetric methods based on AgNO$_3$+fluorescein, AgNO$_3$+K$_2$CrO$_4$ and AgNO$_3$ solutions have been proposed to measure the chloride ion penetration depth in concrete. Both AgNO$_3$+fluorescein and AgNO$_3$+K$_2$CrO$_4$ methods were derived from the methods for analyzing free chloride ions in solutions (Li 2005), as described in Table 6.1. The three methods are described in detail in the following sections.

6.2.1 AgNO$_3$ + Fluorescein Method

In the 1970s, Collepardi et al. (1970 and 1972) developed a colorimetric method to determine the free chloride content in concrete. This method involved a first spray of fluorescein (HF1) solution (an organic weak acid made of 1 g/L of in a 70% solution of ethyl alcohol in water) onto a freshly split surface of a chloride-penetrated concrete, followed by spray of 0.1 mol/L AgNO$_3$ aqueous solution on the same surface. Spray of fluorescein solution results in a yellowish green color due to the presence of F1$^-$ ion in the solution. After spraying the AgNO$_3$ solution, Ag$_2$O and AgCl are precipitated due to the reactions of Ag$^+$ with OH$^-$ and Cl$^-$. Both Ag$^+$ and Cl$^-$ can be absorbed on the surface of AgCl. When a small amount of AgNO$_3$ is sprayed, in other words, Cl$^-$ is excessive; AgCl·Cl$^-$ complex is formed leading to a negative surface, which repels F1$^-$ ion into the solution, giving the solution a yellowish color. When excess AgNO$_3$ is sprayed, Ag$^+$ is absorbed on AgCl, forming AgCl·Ag$^+$ complex with positive surface, which further absorbs F1$^-$ ion from the solution forming AgCl·AgF1 complex, which gives the solution a pink color. The associated reactions, together with the color of the corresponding reaction products, are summarized in Table 6.2. This method has been specified in the Italian Standard 79-28 (Italian Standard 79-28 1978).

Table 6.2 Chemical Reactions Associated with the AgNO₃+Fluorescein Method

Chemical Reactions in Chloride-Containing Zone	*Chemical Reactions in Chloride-Free Zone*
$Ag^+ + Cl^- \rightarrow AgCl$ (white)	
$Ag^+ + OH^- \rightarrow AgOH \rightarrow Ag_2O$ (brown)	$Ag^+ + OH^- \rightarrow AgOH \rightarrow Ag_2O$ (brown)
Excessive Cl^-: $AgCl \cdot Cl^- + Fl^-$ (yellowish green)	
Excessive Ag^+: $AgCl \cdot Ag^+ + Fl^-$ $\rightarrow AgCl \cdot AgFl$ (pink)	

Usually, when this method is applied for measuring chloride ion penetration in concrete, a sufficient amount of $AgNO_3$ solution is sprayed. The chloride-containing zone appears to be dark pink due to the formation of pink $AgCl \cdot AgF1$ on the concrete with a background color of grey. The chloride-free area turns dark brown due to the formation of brown Ag_2O on grey concrete.

6.2.2 AgNO₃+ K₂CrO₄ Method

Since the 1980s, the $AgNO_3$+ K_2CrO_4 method has been used to measure chloride ion penetration in concrete (Maultzsch 1983, 1984; Baroghel-Bouny et al. 1999; RILEM Technical Committee 178 Procedure 2003; Frederiksen 2000; Baroghel-Bouny et al. 2007a, 2007b). In this method, 0.1 mol/L $AgNO_3$ solution with pH in the range of 3– 5 is sprayed onto a freshly split surface of concrete, followed by one hour of natural drying and further spray of a 5 wt.% potassium chromate (K_2CrO_4) solution.

After the K_2CrO_4 solution is sprayed, the chloride-containing zone remains yellow due to the formation of white AgCl and light-yellow K_2CrO_4 solution. A small amount of AgCl (white) and Ag_2CrO_7 (ruby red), may also form, but their amounts are too small to affect the color significantly. Thus, the chloride-containing zone may remain yellowish in most cases. However, the chloride-free zone turns a red-brown color due to the formation of brown Ag_2O + brick red Ag_2CrO_4 and/or ruby red Ag_2CrO_7. A series of chemical reactions taking place in both chloride-containing and chloride-free zones in this colorimetric method are summarized in Figures 6.1 and 6.2.

6.2.3 AgNO₃ Method

$AgNO_3$ method involves spraying only an $AgNO_3$ solution onto a freshly split surface of concrete. The chemical reactions involved in this method are summarized in Table 6.3. Due to precipitation of AgCl and Ag_2O respectively in chloride in chloride-contaminated and free areas, white and brown zones with a clear color change boundary can be observed. The

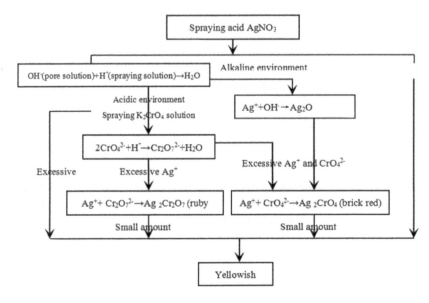

Figure 6.1 Chemical reactions of AgNO$_3$+ K$_2$CrO$_4$ method in chloride-containing zone.

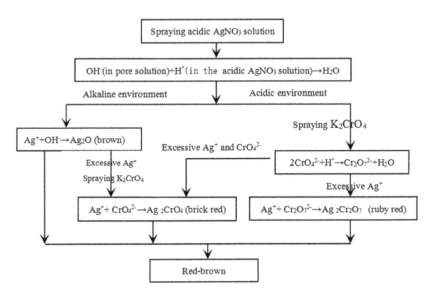

Figure 6.2 Chemical reactions of AgNO$_3$+ K$_2$CrO$_4$ method in chloride-free zone.

depth of the white zone can thus be used to represent the chloride penetration depth. The AgNO$_3$ solution with concentration of 0.1 mol/L was used in this method based on an investigation by Otsuki et al. (1992) of the effect of AgNO$_3$ concentrations from 0.05 to 0.4 mol/L on the brightness of the colors that appeared on the concrete.

Table 6.3 Chemical Reactions in AgNO₃ Method

Chemical Reactions in Chloride-Containing Zone	Chemical Reactions in Chloride-Free Zone
$Ag^+ + Cl^- \rightarrow AgCl$ (white) $Ag^+ + OH^- \rightarrow AgOH \rightarrow Ag_2O$	$Ag^+ + OH^- \rightarrow AgOH \rightarrow Ag_2O$ (brown)

(a) 100%AgCl+0%Ag₂O   (b) 86.1%AgCl+13.9%Ag₂O   (c) 13.9%AgCl+86.1%Ag₂O   (d) 0%AgCl+100%Ag₂O

Figure 6.3 Pictures of AgCl and Ag2O mixtures. (From Yuan et al. 2008.)

The amount of AgCl and Ag₂O formed at the color change boundary from the AgNO₃ method depends on the ratio of Cl⁻ to OH⁻ in the concrete pore solution. Yuan et al. (2008) studied the color of AgCl, Ag₂O, and their mixtures, as shown in Figure 6.3. Pure AgCl is silvery and pure Ag₂O is dark brown. In the presence of a small amount of Ag₂O, the color of the mixture could change gradually from silvery to brown, which will produce a color change boundary.

6.2.4 Comparison of the Three Methods

For all of the three AgNO₃-based colorimetric methods as described above, the color change boundaries are determined by formation of AgCl and AgO, with white and brown colors respectively. The visible color of the color change boundary depends on the relative amount of the two compounds. AgNO₃ method colorimetric mechanism is simplest among the three methods. Both AgNO₃+fluorescein and AgNO₃+K₂CrO₄ methods need longer time to obtain a better color setting at the color change boundary, whereas AgNO₃ method can be conducted more simply and easily than the other two methods. In practice, the color change boundary from the AgNO₃+K₂CrO₄ method is more clearly visible than that from the AgNO₃ method. However, AgNO₃ method is sufficient to show the color change boundary in most cases (Baroghel-Bouny et al. 2007a). Therefore, it is used as the most practical method for measuring chloride penetration in concrete.

The typical colorimetric pictures from the three methods are shown in Figure 6.4, where the color change boundary between chloride-rich and "free" zones can be observed. Obviously, AgNO₃+fluorescein method did

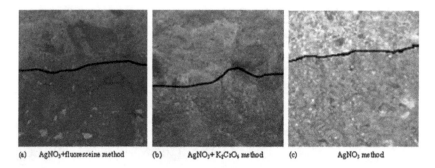

(a)   AgNO₃+fluoresceine method   (b)   AgNO₃+ K₂CrO₄ method   (c)   AgNO₃ method

Figure 6.4 Typical colorimetric pictures from the three methods. (Pictures for a and b from Baroghel-Bouny et al. (2007a); c was taken by the authors.)

not show a clear boundary. Among the three colorimetric methods, color change depths measured with $AgNO_3$+fluorescein method and $AgNO_3$ method are similar. When $AgNO_3$+K_2CrO_4 method is applied, the color change depth was slightly deeper than that from the $AgNO_3$ method (Baroghel-Bouny et al. 2007a). This is because that in the $AgNO_3$+K_2CrO_4 method the solution is acidified to avoid the influence of hydroxyl ions. The similarities and differences among the three methods are summarized in Table 6.4.

The chloride "free" zone indicated by the three different methods is not really chloride free. It may contain a low chloride content, which does not cause significant Cl-related reactions but results in different color compared to the chloride-rich zone, due to the precipitation of Ag_2O. Therefore, the chloride ion penetration depth X_d indicated by the $AgNO_3$-based colorimetric methods does not represent the real chloride ingress depth. However, these measured depths are near to the actual penetration front of chloride (Baroghel-Bouny et al. 2007a, 2007b; He et al. 2008; Chen et al. 1996).

6.2.5 Measurement of Chloride Penetration Depth

NT Build 492 (1999) recommends using slide caliper and a suitable ruler to measure the penetration depths from the center to both edges at intervals of 10 mm after the chloride migration testing (Figure 6.5). Seven depths should be measured and an average of the seven measurements is used for calculating chloride migration coefficients. Special cares should be taken during the penetration depth measurement: (1) If the penetration front to be measured is obviously blocked by aggregate, move the measurement to the nearest front where there is no significant blocking by aggregate or, alternatively, ignore this depth if there are more than five valid depths. (2) If there is a significant defect in the specimen which results in a penetration front much larger than the average, ignore this front as indicative of the penetration depth. (3) To obviate the edge effect due to a non-homogeneous

Table 6.4 Comparisons of Color Change Boundary among the Three Colorimetric Methods

Methods	AgNO$_3$+Fluorescein Method		AgNO$_3$ + K$_2$CrO$_4$ Method		AgNO$_3$ Alone Method
Two colors	Pink and dark brown in the case of excessive Ag$^+$	yellowish green and dark brown in the case of excessive Cl$^-$	Light yellow and brick red in the case of excessive K$_2$CrO$_4$ solution	Light red and brick red in the case of excessive AgNO$_3$ solution	White and brown
Visibility of color change boundary	Low	High	High	Low	Relatively high
Factors influencing visibility of the color change boundary	1. Interval between spraying the two solutions and sprayed amount of second solution. 2. Spraying temperature and humidity also affect optimum interval. 3. Spraying too much K$_2$CrO$_4$ solution could wash precipitated Ag$_2$CrO$_4$ away and result in a dark yellow color (yellow + a little brick-red) in chloride-free area, which affects the clearness of the boundary.				1. Concrete with low alkalinity will cause low visibility or disappearance of color change boundary. 2. Concrete with dark grey color may produce a relatively low visible color change boundary.

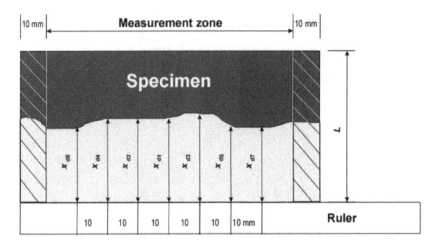

Figure 6.5 Illustration of measurement for chloride change depths NT Build 492 (14).

degree of saturation or possible leakage, do not make any depth measure-ments in the zone within about 10 mm from the edge. Sometimes, due to the interference of aggregate, it is not easy to capture more than five valid penetration depths in practice. In the draft standard (RILEM Technical Committee 178 Procedure 2003), it requires at least four measurements.

Image analysis has also been used to measure the chloride penetration depths (Baroghel-Bouny et al. 2007a, 2007b, Baroghel-Bouny et al. 2002). The average chloride penetration depth X_d is calculated using Equation 6.1:

$$X_d = \frac{S_{Cl^-}}{S} \times L \tag{6.1}$$

where S_{Cl^-} is the area of the chloride-penetrated zone, S and L are the area and the thickness of the specimen respectively.

In the case of image analysis, the subjective feature in the measurement of the penetration front is reduced. However, as follows, there still exist some disadvantages for the image analysis: (1) The image analysis measures average distance between penetrated surface to color change boundary; it can't avoid effect of edge and coarse aggregates of concrete. (2) The image analysis needs calculation of area of the chloride-penetrated zone using cer-tain software, which depends on comparison between colors at two sides of boundary. Different software will produce different errors. (3) Image analysis is influenced by roughness of fractured surface; to its better use, the fractured surface needs dry-cutting. This might help to avoid the errors associated with the coarse surface (Baroghel-Bouny et al. 2007a). (4) Image analysis may be influenced by color variation of concrete and the precipi-tated reaction products on the concrete surface.

Although the caliper and ruler method can't avoid vision error of 0.5 mm, which might produce a small problem on application, according to error evaluation (Tang 1996, He 2010), depth variation of 0.5 mm only caused a small variation of non-steady-state migration coefficient. Therefore, caliper and ruler method can be a proper method to measure penetration depth.

6.3 CHLORIDE CONCENTRATION AT COLOR CHANGE BOUNDARY

Many studies have measured the chloride concentration at the color change boundary; however, the obtained results often controversial contradict to each other. Comparison of these studies is summarized in Table 6.5. As can be seen from Table 6.5 the measured chloride concentrations at the color change boundary are 0.01% by cement based on the AgNO$_3$+fluorescein method (Collepardi 1995), 0.1–0.4% by the mass of cement based on the AgNO$_3$+K$_2$CrO$_4$ method (Baroghel-Bouny et al. 2007a, 2007b), and 0.28%–1.69% by the mass of cement or 0.072-0.714 mol/L in pore solution based on the AgNO$_3$ method (Otsuki et al. 1992; Andrade et al. 1999; Sirivivatnanon and Khatri 1998; Meck and Sirivivatnanon 1999; McPolin et al. 2005; Yuan 2009; He et al. 2008). Several factors during the experiments may explain these discrepancies.

6.3.1 Parameters of Colorimetric Reactions

He (He 2010; He et al. 2011) investigated chloride concentration at the color change boundary of AgNO$_3$ method. They calculated the C_d according to Ag$^+$-Cl$^-$-OH$^+$-H$_2$O system.

If Ag$^+$ reacts with all OH$^-$ and Cl$^-$, chloride concentration ($Cl_{crit-Cl}^-$) for color change of AgCl+Ag$_2$O mixture from white to brown would be expressed as Equation 6.2.

$$C_{crit-Cl}^- = 1.6C_{OH}^- \qquad (6.2)$$

if Ag$^+$ partially reacts with OH$^-$ and Cl$^-$, $C_{crit-Cl}^-$ would be expressed as Equation 6.3.

$$C_{crit-Cl}^- = 0.00695C_{OH}^- + 0.608C_{Ag}^+ V_{Ag}^+ / V_{OH}^-{}_{+Cl}^- \qquad (6.3)$$

where: C_{Ag}^+, and C_{OH}^- are the mole concentrations of Ag$^+$ and OH$^-$.V_{Ag}^+ is the volume of AgNO$_3$ solution added to the NaOH + NaCl solution, $V_{OH}^-{}_{+Cl}^-$ is volume of NaOH +NaCl solution.

The calculated $C_{crit-Cl}^-$ is between 0.01-0.96 mol/L in pore solution (C_{pd}), or 0.011-2.27% by the mass of binder (C_{bd}) (He 2010; He et al. 2011), which is in agreement with results from the literature, i.e., 0.01 to 1.69%

Table 6.5 Summary of Chloride Concentrations at the Color Change Boundary

No.	Method of Introducing Chloride	Indicators	Sampling Method	Method for Measuring Free Chloride	Method for Measuring Total Chloride	Free Chloride	Total Chloride	References
1	Diffusion, migration, and exposure in field	0.1 N AgNO$_3$ and 0.1 N AgNO$_3$+ K$_2$CrO$_4$ solution	Grinding layer by layer	Water extraction method	Acid extraction method	0.1–0.4% by mass of cement	0.2% to 1% by mass of cement	Baroghel-Bouny et al. (2007a) and (2007b)
2	Immersing in NaCl solution	0.1 N AgNO$_3$	Splitting layer by layer	Water extraction method	Fluorescent x-ray	0.15% by cement	0.4–0.5% by cement	Otsuki et al. (1992)
3	Mixing water with NaCl	0.1 N AgNO$_3$+ fluorescence				0.01% by cement		Collepardi (1995)
4	ASTM C1202	0.1 N AgNO$_3$+ fluorescence solution	Not described		Acid extraction method		1.13±1.4% by mass of cement	Andrade et al. (1999)
5	Immersing in NaCl solution	0.1 N AgNO$_3$	Drilling in 2 mm steps	Ion chromatography technique		0.84% to 1.69% by mass of binder		Sirivivatnanon and Khatri (1998)
6	Immersing in NaCl solution	0.1 N AgNO$_3$	Drilling in 2 mm steps	Ion chromatography technique		0.28–1.41% by mass of binder		Meck and Sirivivatnanon (1999)
7	Wetting and drying ponding	0.05-N AgNO$_3$	Drilling at 2 mm steps		UV spectrophotometer method		0.5–1.5% by mass of cement	McPolin et al. (2005)
8	Electrical migration and nature diffusion	0.1 N AgNO$_3$ solution	Grinding layer by layer	Water extraction method	Acid extraction	0.071–0.714 mol/L	0.019%–0.173% by mass of concrete	Yuan (2009)
9	Electrical migration	0.1 and 0.035 N AgNO$_3$ solution	Grinding layer by layer	Water extraction method		0.072–0.142 mol/L		He et al. (2008)

by the mass of binder and 0.071–0.714 mol/L in pore solution (as shown in Table 6.5).

As can be seen from Equations 6.2 and 6.3, concrete alkalinity, the amount and concentration of the sprayed $AgNO_3$ solution, and pore solution volume of concrete can influence chloride concentration at the color change boundary. The smaller the volume of sprayed $AgNO_3$ solution, the lower the $C_{crit-Cl^-}$ is. To obtain a lower $C_{crit-Cl^-}$, $AgNO_3$ solution should be sprayed as little as possible, provided that the surface of the cement-based materials is completely wetted by the $AgNO_3$ solution. The clearest boundary color is observed at 0.1 mol/L $AgNO_3$ solution, which was the suggested use for the indicator. Therefore, a proper sprayed volume of 0.1 mol/L $AgNO_3$ solution should be determined. However, this important point was not taken into account when the $AgNO_3$ method was applied.

As discussed above, concrete alkalinity, sprayed amount, and concentration of $AgNO_3$ solution and pore solution volume of concrete can influence chloride concentration at the color change boundary, measured and calculated C_d change in a large range. Therefore, it is necessary to point out again that a proper sprayed amount of $AgNO_3$ solution with a certain concentration should be determined to obtain a small range of C_d. A smaller C_d range will produce more accurate applications.

6.3.2 Sampling Methods

Different sampling methods and sampling thickness have been used in the literature, which resulted in different chloride concentrations at the color change boundary. From the angle of operation, drilling is better than cutting or grinding. This is because drilling is carried out in the central region of the specimens and does not consider effect of specimen edge. When considering homogeneity of sampling, especially at color change boundary, cutting or grinding is better than drilling. This is because drilling is just from one or several locations, which may not be very representative. This may be observed from Figure 6.6. However, cutting or grinding is from the entire surface within central region, which is more representative. As shown in Table 6.5, the chloride concentration range from grinding or cutting is much smaller than that from drilling. It seems that cutting or grinding is more suitable for sampling.

6.3.3 Methods for Free Chloride Measurement

Many methods, such as pore solution expression, water extraction method (water soluble chloride), alkaline solution extraction method, nuclear magnetic resonance (NMR) etc. have been used to determine the free chloride concentration. Both ion chromatography and nuclear magnetic resonance techniques need special and costly equipment. Among these methods, the results from pore solution expression are regarded as being close to the real free

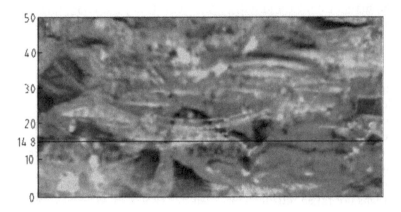

Figure 6.6 Pictures of the color change boundary of concrete. (From He 2010.)

chloride concentration (Arya and Newman 1990; Barneyback and Diamond 1981, 1986). However, this method needs special expression apparatus, and it is difficult to obtain enough pore solution from concrete with low water-to-cement ratio and maybe produce more chloride concentration than free chloride in bulk pore solution (He et al. 2016). The water extraction method has often been used for analyzing chloride content of the pore solution of harden concrete owing to its simple operation (Arya and Newman 1990; Arya et al. 1987; Glass et al. 1996; Haque and Kayyali 1995). He (2010) confirmed that the water extraction method can be used as an effective method to analyze the chloride content at the color change boundary. However, it is important to control experimental parameters, because experimental results derived from different extraction parameters may produce large difference. Many researchers (Baroghel-Bouny et al. 2007a, 2007b; Otsuki et al. 1992; Yuan 2009; He et al. 2008) used different water extraction parameters for measuring free chloride concentration at the color change boundary, which is possibly partly responsible for the C_d with large range.

6.3.3.1 Pore Solution Expression Method

The expression method has been widely used for analysis of the ion concentration in a pore solution of cement-based materials (Page et al. 1991; Page and Vennesland 1983; Yonezawa et al. 1988; Arya et al. 1990). Studies using the expression method (Barneyback and Diamond 1986; Buckley et al. 2007; Mohammed and Hamada 2003) have mainly focused on its feasibility, while some studies have focused on the reliability of the results, which is assumed to be very close to the actual ion concentration in the pore solution.

During the studies in pore solution of cement-based materials by pore solution expression method, the phenomenon of "chloride concentrate" and chloride concentration index have been proposed (Someya et al. 1989; Nagataki et al. 1993; Ishida et al. 2008). As discussed in Chapter 3, the pore structure and surface properties of cement hydrate (zeta potential) mainly affects the results of chloride concentration of pore solution obtained by pore solution expression. The vibration of pore structure and chloride distribution within cement paste during pore solution expression are applied to calculate the chloride concentration of pore solution, as shown in Section 3.5.4.

Based on results from Larsen et al (Larsen 1998; Viallis-Terrisse et al. 2001), surface potential of cement hydrate is always higher than zero when the alkaline cation concentration in the pore solution is large enough. Indeed, there is usually sufficient alkaline cation concentration in the pore solution of cement-based materials, namely, surface potential >0 in most situations; this explains why chloride concentration index has been reported to be higher than 1 in most publications (Yuan 2009; Li et al. 2013; Someya et al. 1989; Nagataki et al. 1993; Ishida et al. 2008; Sugiyama et al. 2003; He 2010). In addition, many factors influencing electric double layer (EDL) (Li et al. 2013; He 2010; He et al. 2016), such as concentration of soaking solution, soaking time, curing time, soaking temperature, expression pressure, etc. will change the tested results. Therefore, pore expression method has a variation in the results.

6.3.3.2 Water Extraction Method

The chloride ion concentration measured by the water extraction method may overestimate the chloride ion concentration of the pore solution, since chemically and physically bound chloride ions may be released when cement-based materials are exposed to large fraction of water (Chaussadent and Arliguie 1999, Shi et al. 2017 and Tang and Nilsson 2001). Chaussadent et al. (1999) showed that chemically bound chloride has been partially dissolved after being added to water for more than three minutes. Thus, they suggested an extraction time of three minutes for the water extraction method. Destabilization of Friedel's salt due to exposure to a large fraction of solution was also reported by others (Birnin-Yauri and Glasser 1998; Suryavanshi and Swamy 1996), which was ascribed to the lowering of pH below 12 (Birnin-Yauri and Glasser 1998). These observations reflected the fact that the water-extracted chloride ion concentration measured by most researchers was larger than the free chloride ion concentration determined by pore solution expression method (Haque and Kayyali 1995) or equilibrium method (He et al. 2016). Especially in the case of low free chloride ion concentration, the concentration of the water extracted chloride can be over seven-fold higher (Arya et al. 1987).

Arya et al. (Arya et al. 1987; Muralidharan et al. 2005; Pavlík 2000; He 2010) reported that water-extracted chloride amount was influenced by the fineness of powder, extraction temperature, stirring time, water/solid ratio, and exposure time for cement-based materials under similar conditions, such as methods for introducing chloride, cations of chloride salt, free chloride ion concentration, and binder types. He et al. (2018)

calculated water-extracted chloride (C_{pws}) for the understanding or even accurate determination of free chloride. The detailed description on the calculation can be seen in a publication by He et al. (2018). Calculated results indicated that with an increase of powder fineness, C_{pws} increases until a fixed value (approximately after the fineness of powder is higher than 1um) is obtained when the water/solid ratio changes from 0.4 to 100. The results also show that the measured C_{pws} increases with the increase of water/solid ratio in form of a power. Similar trends have also been reported by Pavlík (2000). This means water-extracted chloride amount increase with increase of water/solid ratio due to de-binding of chemically bound chloride. At the same time, increase of the release of chemically bound chloride can occur with increasing extraction temperature.

Based on the above discussion, it can be found that chloride tested by the expression method contains free chloride and part-physically bound chloride, and chloride tested by water extraction method may consist of free, physically bound (total or most), and part-chemically bound chloride. Depending on experimental parameters, chloride amount tested by the expression and water extraction method can vary in a relatively large range. Therefore, when using these two methods to test chloride concentration at the color change boundary, it may produce an unreliable assessment.

6.3.4 Spraying of AgNO$_3$ Solution and Representative Value of C_d

6.3.4.1 Spraying Method and Amount of AgNO$_3$ Solution

It can be seen from Equation (6.3) that spraying a lower amount of AgNO$_3$ solution can get lower C_d. However, the amount of spraying of silver nitrate solution should not be infinitely small, because it requires a certain amount of silver nitrate solution to entirely wet whole section of concrete. The storage of silver nitrate solution is different between horizontal placement and vertical placement of cement-based materials sections. When the section is placed horizontally, the excess silver nitrate solution will remain in the section, and when the section is placed vertically, the excess silver nitrate solution overflows from the section under the gravity. Therefore, the section placement method will lead to the different absorption amount of silver nitrate solution by the section of cement-based material, which leads to a different amount of silver nitrate in the reaction.

He (2010) found that when the section was placed vertically, the color change depths measured by 1.5ml spraying amount of 0.1mol/L AgNO$_3$ solution were larger than those by 3 and 4.5ml spraying amount of 0.1 mol/L AgNO$_3$ solution, typically as shown in Figure 6.7. When the section was placed horizontally, there were very small differences among the color change depths measured by 1.5, 3, and 4.5 ml spraying amount of 0.1 mol/L AgNO$_3$ solution, respectively. Therefore, He (2010) suggested that

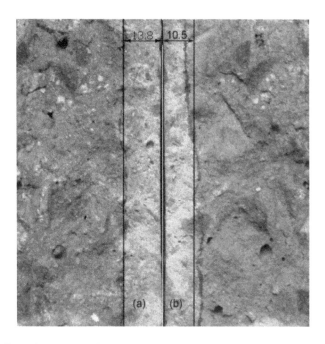

Figure 6.7 Typical pictures of colorimetric measurement from different volume of 0.1 mol/L AgNO₃ solution. (a) Sprayed 1.5 ml AgNO₃ solution for 50cm² area; (b) sprayed 3 ml AgNO₃ solution 50cm² area. (From He 2010.)

when the $AgNO_3$ colorimetric method was conducted, the section of concrete should be placed horizontally. Considering completely wet section of concrete, a $0.3\pm0.06L/m^2$ spraying amount of 0.1 mol/L $AgNO_3$ solution was recommended by He (2010).

6.3.4.2 Representative Value of C_d

Considering different surfaces of concrete may have different absorption to the $AgNO_3$, the measured C_d may be different even with spraying of constant amount of 0.1 N $AgNO_3$ solution. Moreover, spraying the same amount of $AgNO_3$ solution is very difficult to control. He (2010) measured large numbers of C_d values based on spraying method and amount of $AgNO_3$ solution given in Section 6.3.4.1; the results are listed in Table 6.6. It can be seen from Table 6.6 that mineral admixture types and water binder ratio has effect on measured C_d, even using same spraying amount and method of solution. It can be seen from Figure 6.8 that with increase of w/b, C_d decrease and the decreasing degree gradually lower. Based on Table 6.6 and Figure 6.8, representative values of C_d are listed in Table 6.7. Those representative values of C_d only produce <5% error of non-steady-state migration coefficient, which can be seen in Section 6.3.6.

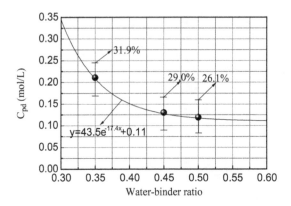

Figure 6.8 Relationship between water-binder ratio and Cd. (From He 2010.)

Table 6.6 Statistical Results of All Measured Chloride Concentrations at Color Change Boundary

Specimens	Range (mol/L)	Average C_{pd} (mol/L)	COV (%)	Numbers of Data
With SF	0.075–0.204	0.132	27.4	24
With SF+0.35W/B	0.075–0.204	0.142	27.8	16
With SF+0.5 or 0.45 W/B	0.08–0.141	0.111	23.1	8
With FA	0.055–0.331	0.178	42.9	52
With FA +0.35W/B	0.072–0.331	0.227	28.0	28
With FA +0.5 or 0.45 W/B	0.055–0.195	0.120	33.1	24
With SL	0.061–0.345	0.184	42.1	52
With SL +0.35W/B	0.084–0.345	0.235	28.4	28
With SL +0.5 or 0.45 W/B	0.061–0.192	0.124	25.7	24
OPC	0.085–0.265	0.183	35.2	14
OPC+0.35W/B	0.15–0.265	0.226	16.9	8
OPC+0.5 or 0.45 W/B	0.085–0.181	0.125	32.9	6
All	0.055–0.345	0.173	41.8	142
0.35W/B	0.072–0.345	0.213	31.8	80
0.5 or 0.45 W/B	0.055–0.195	0.121	28.8	62

He (2010).

Note: SF is silica fume, FA is fly ash, SL is slag powder.

Table 6.7 Determination of Chloride Concentration Range at the Color Change
Boundary in the Case of Saturated Cement-Based Materials

W/B of Specimens	0.5	0.45	0.35
C_d for specimens without SF	0.1 mol/L	0.15 mol/L	0.2 mol/L
C_d for specimens with SF	0.1 mol/L		

6.4 APPLICATION OF COLORIMETRIC METHOD TO DETERMINE CHLORIDE DIFFUSION/MIGRATION COEFFICIENT

6.4.1 Measurement of Non-Steady-State Chloride Diffusion

Non-steady-state chloride diffusion can be described by Fick's second law, whose analytical solution is expressed as Equation 6.4.

$$C(x,t) = C_s \left[1 - \text{erf}\left(\frac{X_d}{2\sqrt{D_{app} \cdot t}} \right) \right] \tag{6.4}$$

where $C(x,t)$ is chloride concentration at a certain depth X_d and time t; C_s is surface chloride concentration; erf is error function; X_d is chloride penetration depth; t is penetration time; D_{app} is chloride apparent diffusion coefficient.

Fick's second law is the theoretical base of NT Build 443 (1995). In NT Build 443, C_S and D_{app} are obtained by fitting the experimental chloride profile to Equation 6.4 by means of a non-linear regression analysis using the least square method.

6.4.1.1 Measurement of Chloride Penetration Kinetics

Equation 6.4 can be written as follows:

$$X_d = 2\text{erf}^{-1}\left(1 - \frac{C_d}{C_s} \right)\sqrt{D_{app}t} \tag{6.5}$$

where erf⁻¹ is inverse error function, C_d is chloride concentration at X_d.

If C_S, C_d and D_{app} change in a small range, then Equation 6.6 is obtained:

$$2\text{erf}^{-1}\left(1 - \frac{C_d}{C_s} \right)\sqrt{D_{app}} = B \tag{6.6}$$

Equation 6.5 can be written as Equation 6.7:

$$X_d = B\sqrt{t} \tag{6.7}$$

Equation 6.7 can be used to describe the chloride penetration kinetics. Baroghel-Bouny et al. (Baroghel-Bouny et al. 2007a and 2007b; He 2010) found good linear correlation between X_d and $\sqrt{t}$ for certain concrete. This means that the coefficient B in Equation 6.7 is nearly a constant for certain concrete. Based on this equation, X_d can be easily assessed. The obtained kinetics can also be used to monitor the residual life prediction of existing concrete structures.

6.4.1.2 Measurement of Apparent Chloride Diffusion Coefficient

Consider no binding of chloride, the apparent D_{app} in Equation 6.4 could be calculated according to the total chloride profiles (Tang 1996), where both free C_S and C_d should be fitted by Equation 6.4.

As chloride profile grinding is time consuming and laborious, an $AgNO_3$-based colorimetric method is conducted to measure the C_d and X_d, according to the modified Equation 6.8 as follows:

$$D_{app} = \left(\frac{X_d}{2\,\mathrm{erf}^{-1}\left(1 - \dfrac{C_d}{C_S}\right)\sqrt{t}} \right)^2 \tag{6.8}$$

Baroghel-Bouny et al (Baroghel-Bouny et al. 2007a) compared the results between the apparent diffusion coefficients (D_N) obtained from the profile method (NT Build 443) and those (D_A) obtained by Equation 6.8, in which penetration depths were determined from the $AgNO_3$ colorimetric method and $C_d/C_S = 0.14$ was used based on a previous publication (Tang 1996b). However, a good accord of D_A with D_N was not observed. Chiang and Yang (2007) took chloride concentration in contact solution as C_S ($C_S = 0.53$ mol/L), associated with $C_d = 0.07$ mol/L (Tang 1996); they found that there was a good linear relationship between D_A and D_N (see Figure 6.9). He's investigation results (2010) may explain above contradiction. Based on proper $AgNO_3$ sprayed amount and thus a small measured C_d, He (2010) investigated relationship between D_N and D_A. Relationships between D_N and D_A are given in Figure 6.10. It can be seen from Figure 6.10 that when using chloride concentration in soaking solution as surface chloride concentration, only part of D_A and D_N are similar after 90-day exposure, and all of D_A and D_N are similar after 300-day exposure. It can be also observed from Figure 6.10 that when using chloride concentration in surface layer (2 mm) as surface chloride

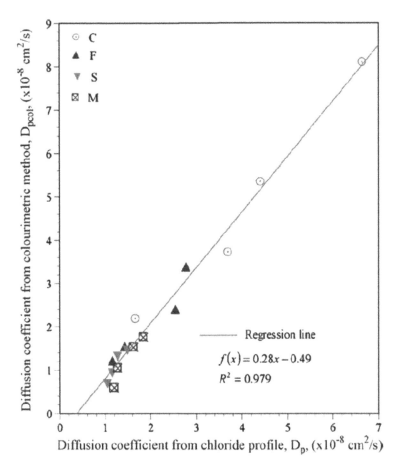

Figure 6.9 Relationship between apparent diffusion coefficients measured by the profile method and AgNO₃ colorimetric method. (From Chiang and Yang, 2007.)

concentration, all of D_A and D_N are very similar after 90-day or 300-day exposure. This means that once the chloride color change depth is measured using AgNO₃ colorimetric method, the D_A can be quickly calculated, which can be a quick and relatively accurate estimation of D_N from the profile method in the case of measuring chloride concentration in surface layer. Therefore, depending on determination of C_S, AgNO₃ colorimetric method can rapidly and relatively accurately measure chloride apparent diffusion coefficient.

6.4.2 Measurement of Non-Steady-State Electrical Migration

Many methods have been developed and proposed to measure and calculate the migration of chloride ion in concrete under both steady- and

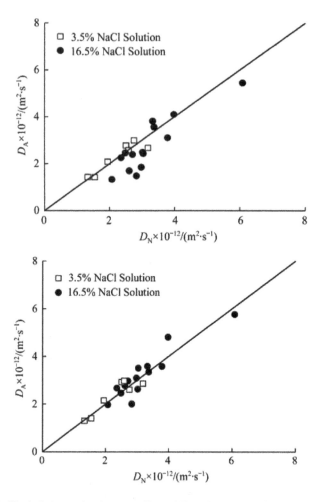

Figure 6.10 (Top): Relationship between D_A and D_N when taking chloride concentration in the immersion solution as the Cs. (Bottom): Relationship between D_A and D_N values when taking chloride concentration in the 2mm surface layer as the Cs. (Data from He 2010.)

non-steady-state conditions. A comparison of these testing methods (Shi et al. 2007) indicates that NT Build 492 is the most appropriate method for accelerated chloride migration testing under non-steady-state conditions (NT Build 492 1999). In this method, a sole 0.1 mol/L AgNO$_3$ solution is used to measure the chloride penetration depth after the non-steady-state migration testing. The non-steady-state chloride migration coefficients are calculated according to Equations 6.9 and 6.10:

$$D_{nssm} = \frac{RT}{ZFE} \cdot \frac{xd - \alpha\sqrt{xd}}{t} \qquad (6.9)$$

$$\alpha = 2\sqrt{\frac{RT}{ZFE}} \cdot \text{erf}^{-1}\left(1-2\frac{C_d}{C_s}\right)$$ (6.10)

where, D_{nssm}: non-steady-state migration coefficient, m²/s; Z: absolute value of ion valence, for chloride, $z = 1$; F: Faraday constant, $F = 9.648 \times 104$ J/ (V·mol); U: absolute value of the applied voltage, V; R: gas constant, $R = 8.314$ J/ (K·mol); T: average value of the initial and final temperatures in the anolyte solution, K; L: thickness of the specimen, m; t: test duration, seconds; erf⁻¹: inverse of error function; C_0: chloride concentration in the catholyte solution, $C_0 \approx 2$ mol/L. x_d and C_d are chloride penetration depth and chloride content at which the color changes in the case of AgNO₃ colorimetric measurement.

The non-steady-state migration testing method, or NT Build 492, proposed by Tang (1996), assumes a chloride concentration of 0.07 mol/L at the color change boundary of ordinary Portland cement specimens. This concentration was based on the research carried out by Otsuki et al. (Tang and Nilsson 1992). However, from above description, the chloride content at the color change ranges from 0.071-0.714 mol/L in pore solution, which is significantly different from the number assumed by Tang (1996). This C_d range is significantly different from 0.07 mol/L used in NT Build 492.

6.4.2.1 Effect of C_d on D_{nssm} Error

He et al. (2012) evaluated systematic errors of D_{nssm} caused by C_d. Equation 6.11, as given in Tang (1996), was used for estimating systematic errors caused by C_d. If C_d changes within calculated range (0.03–1.02 mol/L), systematic errors caused by C_d are plotted in Figure 6.11. It can be seen from Figure 6.11 that, although the X_d is more than 30 mm, the error is

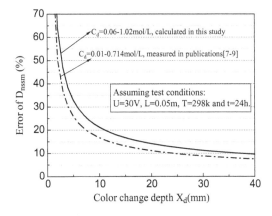

Figure 6.11 Estimated systematic errors caused by the C_d. (From He et al. 2012.)

still beyond 10%. In the case of X_d is less 10 mm, the error is more than 20%. When the X_d is 5 mm, the error is beyond 40%. It can be also seen in Figure 6.11, the smaller the X_d is, the larger the error of D_{nssm} is. For high performance concrete, X_d may reach 5–10 mm; the error derived from the C_d may be relatively large.

$$\text{COV}(\Delta\alpha)=\frac{\Delta D_{nssm}(\Delta\alpha)}{D_{nssm}}\times100 = \left|\frac{\alpha}{\alpha-\sqrt{x_d}}\right|\times\left|\frac{\Delta\alpha}{\alpha}\right| \qquad (6.11)$$

6.4.2.2 Error of D_{nssm} Based on Controlled Amount of Sprayed AgNO$_3$ Solution

He et al. (2012) also calculated errors of D_{nssm} caused by C_d measured based on the established AgNO$_3$ spraying method, as stated in Section 6.3.4, and the results are plotted in Figure 6.12. It can be seen from Figure 6.12 that when taking 0.07 mol/L AgNO$_3$ solution as C_d to use for calculating D_{nssm}, even C_d has a relatively small range; the error still gets 20% at about X_d = 5 mm. This is still bad news for high performance concrete. When C_d = 0.2 mol/L is used for calculating D_{nssm}, the error can be controlled at below 5%, even X_d < 5 mm. The above discussions indicate that C_d = 0.07 mol/L may be not suitable for the calculation of D_{nssm} using AgNO$_3$ colorimetric method. The established AgNO$_3$ spraying method and C_d = 0.2 mol/L can be reasonable for measuring and calculating D_{nssm}. Therefore, it is necessary for controlling the amount of sprayed AgNO$_3$ solution to get smaller error of D_{nssm}.

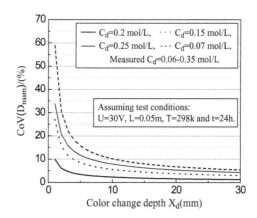

Figure 6.12 Error of D$_{nssm}$ calculated according to C$_d$ measured based on the established AgNO$_3$ spraying procedure. (From He et al. 2012.)

6.4.3 Evaluation for Corrosion Risk of Steel in Concrete

Evaluation for corrosion risk of steel in concrete is based on comparisons between threshold chloride concentration for steel corrosion (C_{crit}) and chloride concentration at color change boundary. The reported threshold chloride concentration with respect to corrosion risk varies greatly (Thomas and Matthews 2004; Alonso et al. 2000). The Building Research Establishment (Everett and Treadaway) has proposed a classification for assessing the risk of corrosion in terms of acid-soluble (or total) chloride contents by the mass of cement: low, less than 0.4%; medium, 0.4–1.0%; and high, greater than 1%. Based on measured range of C_d, Meck et al. (Baroghel-Bouny et al. 2007b; Meck and Sirivivatnanon 1999) suggested that owing to the fact that the range of C_d is generally lower than C_{crit}, evaluation for corrosion risk of steel in concrete can be operated using AgNO₃ colorimetric method.

6.5 CHLORIDE ION TYPES IN C_d AND ABSENCE OF COLOR CHANGE BOUNDARY

As stated above, when spraying AgNO₃ solution, both OH⁻ and Cl⁻ can react with Ag⁺. When Cl⁻ concentration in pore solution decreases due to the reaction, chloride in the diffuse layer will be released to react with excess Ag⁺. When OH⁻ concentration in pore solution decreases due to the reaction, pH of pore solution will gradually decrease and thus maybe lead to the release of chemically bound chloride into pore solution, which will be reacted by excess Ag⁺. Thus, C_d may be affected by free chloride, as well as physically and chemically bound chloride. This means that when AgNO₃ increases, measured C_d are more affected by the physically and chemically bound chloride, which produces variation of C_d of the pore solution. Therefore, controlling spraying amount of AgNO₃ is very important to get small range of C_d.

C_d can vary within a large range, depending on spraying amount of AgNO₃ solution and alkalinity of concrete. When the alkalinity is high and penetrated Cl⁻ concentration is very low, absence of color change boundary may happen. In this situation, the color change boundary can appear when using lower concentration of AgNO₃ solution (0.035 mol/L) (He 2010). In addition, carbonated concrete has low pore solution alkalinity; after spraying AgNO₃ solution, a small amount of brown Ag₂O forms, and absence of color change boundary will be obtained in the case of low Cl⁻ concentration. Therefore, when concentration and spraying amount of AgNO₃ solution are higher and amount of penetrated chloride is small or concrete section color is dark, color change boundary may not appear. This situation is bad news for application of AgNO₃ colorimetric method in field concrete with a low alkalinity and high-performance concrete with small penetration amount of chloride.

6.6 DEPTH DEPENDENCE OF CHLORIDE DIFFUSION COEFFICIENT BASED ON AGNO₃ COLORIMETRIC METHOD

In fact, $AgNO_3$ colorimetric method only measures one penetration depth and chloride concentration at this penetration depth. A question will be proposed whether diffusion coefficient based one depth and the corresponding concentration can be similar with that based on the profile method in NT Build 443. Tang (1996) pointed out that there is a depth dependence of D_{nssm}, namely, an increased D_{nssm} will be obtained with the increase of chloride penetration depth. He (2010) investigated depth dependence of chloride diffusion coefficient. Results indicated that depending on selection of surface chloride concentration, the chloride diffusion coefficient was highly dependent on the chloride penetration depth. Chloride diffusion coefficient near concrete surface is quite different from the inner concrete. Thus, when using $AgNO_3$ colorimetric method to measure chloride migration coefficient, the depth dependence of chloride migration coefficient will be an issue worthy of consideration and further investigation. Associated with significant effect of surface chloride concentration C_S (He 2010), it may be more reasonable to consider combined effect C_d, C_S, and depth dependence when using $AgNO_3$ colorimetric method.

6.7 SUMMARY

This paper reviews three $AgNO_3$-based colorimetric methods, factors affecting the color change and chloride ion concentrations at the color change boundary, and their applications. Some important aspects can be summarized as follows:

$AgNO_3$-based colorimetric methods are fast and easy to perform. They have been widely applied to measure the chloride penetration depth in the field and the laboratory. Among the three methods, the $AgNO_3$ method can be conducted more rapidly, simply, and easily than $AgNO_3$+ K_2CrO_4 or $AgNO_3$+fluorescein method; the three methods give similar measured results.

The color change boundary shown by the colorimetric method is not a real boundary between chloride-free zone and chloride-penetrated zone. The boundary has a certain chloride concentration owing to OH⁻ ions in pore solution of concrete. The measurement of the chloride ion penetration depth can be carried out on specimens by caliper and ruler method or by means of analyzing the digitized image of the specimen. Caliper and ruler method can be a proper method to measure penetration depth.

Analysis of published results indicate that the chloride concentrations at the color change boundary is 0.01% by the mass of cement for $AgNO_3$+fluorescein method, 0.1–0.4% by the mass of cement for $AgNO_3$ +K_2CrO_4 method, and 0.28%–1.69% by the mass of cement or 0.072–0.714 mol/L for the $AgNO_3$ method. Many factors such as concrete alkalinity, sprayed volume and concentration of $AgNO_3$ solution, pore solution volume of concrete, sampling method, and methods used for measuring free chloride in concrete can be responsible for the high variability. Among all factors, sprayed volume and concentration of $AgNO_3$ solution are main ones.

The smaller the volume of sprayed $AgNO_3$ solution with a certain concentration, the lower the C_d is. 0.1 mol/L $AgNO_3$ solution has the most clear color of boundary. To obtain a lower C_d, a proper sprayed volume of 0.1mol/L $AgNO_3$ solution should be determined. Based on small range of C_d, the method can be potentially to be used as a useful tool within the framework of evaluating chloride penetration in reinforced concrete structure in chloride environments.

If the volume of sprayed $AgNO_3$ solution isn't controlled, C_d variation may produces 40% error of D_{nssm} when X_d is about 5mm. This is bad news for high-performance concrete. When the volume of sprayed $AgNO_3$ solution is controlled, in the case of 0.2 mol/L used for calculating D_{nssm}, less than 5% error of D_{nssm} caused by C_d is obtained; even X_d is less than 5 mm. C_d = 0.07 mol/L is not suitable to be used for calculating D_{nssm} in the $AgNO_3$ colorimetric method. The established $AgNO_3$ spraying method by He (2010) and C_d = 0.2 mol/L is more reasonable for measuring and calculating D_{nssm}.

Depending on determination of C_S, $AgNO_3$ colorimetric method can rapidly and relatively accurately measure chloride apparent diffusion coefficient. This means that once the chloride color change depth is measured using $AgNO_3$ colorimetric method, the D_A can be quickly calculated, which can be a quick and relatively accurate estimation of D_N from the profile method.

When concentration and spraying of $AgNO_3$ solution is high and amount of penetrated chloride is low, color change boundary may not be visible. Namely, when concrete was carbonated and amount of penetrated chloride is low, color change boundary may also be invisible. Thus, this method is not suitable for field concrete with a low alkalinity and high-performance concrete with very low penetrated chloride amount, such as silica fume concrete.

Due to depth dependence of chloride migration coefficient, when using $AgNO_3$ colorimetric method to measure chloride migration coefficient, the depth dependence of chloride migration coefficient will be an issue worthy of consideration and further investigation. Associated with significant effects of C_S, it may be more reasonable to consider combined effect C_d, C_S, and depth dependence when using $AgNO_3$ colorimetric method.

REFERENCES

Alonso C, Andrade C, Castellote M, Castro P. Chloride threshold values to depassivate reinforcing bars embedded in a standardized OPC mortar. *Cement and Concrete Research*. 2000, 30:1047–1055.

Andrade C, Castellote M, Alonso C, Gonzalez C. Relation between colorimetric chloride penetration depth and charge passed in migration tests of the type of standard ASTM C1202-91. *Cement and Concrete Research*. 1999, 29:417–421.

Arya C, Buenfeld NR, Newman JB. Assessment of simple methods of determining the free chloride ion content of cement paste. *Cement and Concrete Research*. 1987, 17:907–918.

Arya C, Buenfeld NR, Newman JB. Factors influencing chloride binding in concrete. *Cement and Concrete Research*. 1990, 20:291–300.

Arya C, Newman JB. An assessment of four methods of determining the free chloride content of concrete. *Materials and Structures*. 1990, 23:319–330.

Barneyback RS, Diamond S. Expression and analysis of pore fluids from hardened cement pastes and mortars. *Cement and Concrete Research*. 1981, 11:279–285.

Barneyback RS, Diamond S. Expression and analysis of pore fluids from hardened cement pastes and mortars. *Cement and Concrete Research*. 1986, 16:760–770.

Baroghel-Bouny V, Arnaud S, Henry D, Carcasse`s M, Que´nard D. Ageing of concretes in natural environments: Experimentation for the 21st century. III—Durability properties of concretes measured on laboratory specimens. *Bulletin des Laboratoires des Ponts et Chaussees*. 2002:13–59.

Baroghel-Bouny V, Belin P, Maultzsch M, Henry D. $AgNO_3$ spray tests: Advantages, weaknesses, and various applications to quantify chloride ingress into concrete. Part: Non-steady-state diffusion tests and exposure to natural conditions. *Materials and Structures*. 2007a, 40:759–781.

Baroghel-Bouny V, Belin P, Maultzsch M, Henry D. $AgNO_3$ spray tests: Advantages, weaknesses, and various applications to quantify chloride ingress into concrete. Part: Non-steady-state migration tests and chloride diffusion coefficients. *Materials and Structures*. 2007b, 40:783–799.

Baroghel-Bouny V, Chaussadent T, Henry D. Testing and modeling chloride penetration into concrete—First round robin test. RILEM Technical Committee 178 Report, LCPC; 1999, 15.

Birnin-Yauri UA, Glasser FP. Friedel's salt, Ca2Al(OH)6(Cl,OH) 2H2O: Its solid solutions and their role in chloride binding. *Cement and Concrete Research*. 1998, 28:1713–1723.

Buckley LJ, Carter MA, Wilson MA, Scantlebury JD. Methods of obtaining pore solution from cement pastes and mortars for chloride analysis. *Cement and Concrete Research*. 2007, 37:1544–1550.

Chaussadent T, Arliguie G. AFREM test procedures concerning chlorides in concrete: Extraction and titration methods. *Materials and Structures*. 1999, 32:230–234.

Chen G, Li W, Wang P. Penetration depth and concentration distraction of chloride ions into cement mortar (in Chinese). *Journal of Tong Ji University*. 1996, 24:19–24.

Chiang CT, Yang CC. Relation between the diffusion characteristic of concrete from salt ponding test and accelerated chloride migration test. *Materials Chemistry and Physics*. 2007, 106:240–246.

Collepardi M, Marcialis A, Turriziani R. Kinetics of penetration of chloride ions into concrete (in Italian). *Il Cemento*. 1970, 67:157–164.

Collepardi M, Marcialis A, Turriziani R. Penetration of chloride ions in cement pastes and in concretes. *Journal of the American Ceramic Society*. 1972, 55:534–535.

Collepardi M. Quick method to determine free and bound chlorides in concrete. In: Nilsson LO, Ollivier JP (eds.). Proceedings of the International RILEM Workshop-Chloride Penetration Into Concrete, Paris, 1995, 10–16.

Everett LM, Treadaway KWJ. Deterioration due to corrosion in reinforced concrete. Information Paper IP 12/80, Building Research Establishment, UK, 19.

Frederiksen JM. Testing chloride in structures—An essential part of investigating the performance of reinforced concrete structures. Proceedings of COST 521 Workshop, Belfast, UK, 2000.

Glass GK, Wang Y, Buenfeld NR. An investigation of experimental methods used to determine free and total chloride contents. *Cement and Concrete Research*. 1996, 26:1443–1449.

Haque MN, Kayyali OA. Free and water soluble chloride in concrete. *Cement and Concrete Reseach*. 1995, 25:531–542.

He F. Measurement of chloride migration in cement-based materials using AgNO₃ colorimetric method. Ph.D. dissertation. School of Civil Engineering and Architecture, Central South University, Changsha, China, 2010.

He F, Shi C, Chen C, An X, Tong B. Error analysis for measurement of non-steady state chloride migration coefficient in concrete. *Journal of the Chinese Ceramic Society*. 2012, 40:20–26.

He F, Shi C, Hu X, Wang R, Shi Z, Li Q, Li P, An X. Calculation of chloride ion concentration in expressed pore solution of cement-based materials exposed to a chloride salt solution. *Cement and Concrete Research*. 2016, 89:168–176.

He F, Shi C, Yuan Q, An X, Tong B. Calculation of chloride concentration at color change boundary of AgNO₃ colorimetric measurement. *Cement and Concrete Research*. 2011, 41:1095–1103.

He F, Shi C, Yuan Q, et al. A Study on factors influencing chloride concentration at color change boundary of concrete using AgNO₃ colorimetric method (in Chinese). *Journal of The Chinese Ceramic Society*. 2008, 36:890–895.

He F, Wang R, Shi C, Zhang R, Shi Z, Zhang D. Effect of bound chloride on extraction of water soluble chloride in cement-based materials exposed to a chloride salt solution. *Construction and Building Materials*. 2018, 160:223–232.

Ishida T, Miyahara S, Maruya T. Chloride binding capacity of mortars made with various Portland cements and mineral admixtures. *Journal of Advanced Concrete Technology*. 2008, 6:287–301.

Italian Standard 79–28. Determination of the chloride. Rome; 1978.

Larsen CK. Chloride binding in concrete-effect of surrounding environment and concrete composition. Ph.D. thesis. The Norwegian University of Science and Technology, Norway; 1998.

Li KA. *Analytical Chemistry*. Peking University Press; 2005.

Li Q, Shi C, He F, Xu S, Hu X. Investigation on factors influencing free chloride ion condensation in cement-based materials (in Chinese). *Journal of the Chinese Ceramic Society*. 2013, 41:320–327.

Measurement of the colorimetric front, RILEM Technical Committee 178 Procedure. Preliminary draft; 2003.

Maultzsch M. Concrete related effects on chloride diffusion (in German). Contribution to Int. Coll. Chloride Corrosion, Vienna, February 22–23, 1983, BAMAmts-und. Mitteilungsblatt 13:387–389.

Maultzsch M. Effects on cement pastes and concrete of chloride solution impact (in German). *Material und Technik* (CH). 1984, 12:83–90.

McPolin D, Basheer PAM, Long AE, Grattan KTV, Sun T. Obtaining progressive chloride profiles in cementitious materials. *Construction and Building Materials*. 2005, 19:666–673.

Meck E, Sirivivatnanon V. Field indicator of chloride penetration depth. *Cement and Concrete Research*. 1999, 33:1113–1117.

Mohammed TU, Hamada H. Relationship between free chloride and total chloride contents in concrete. *Cement and Concrete Research*. 2003, 33:1487–1490.

Muralidharan S, Vedalakshmi R, Saraswathi V, Joseph J, Palaniswamy N. Studies on the aspects of chloride ion determination in different types of concrete under macro-cell corrosion conditions. *Building and Environment*. 2005, 40:1275–1281.

Nagataki S, Otsuki N, Wee TH, Nakashita K. Condensation of chloride Ion in hardened cement matrix materials and on embedded steel bars. *ACI Materials Journal*. 1993, 90:323–332.

NT Build 443. *Nordtest Method, Accelerated Chloride Penetration*. Espoo, Finland; 1995.

NT Build 492. *Nordtest Method, Concrete, Mortar and Cement-based Repair Materials: Chloride Migration Coefficient from Non-steady-state Migration Experiments*. Espoo, Finland; 1999.

Otsuki N, Nagataki S, Nakashita K. Evaluation of AgNO3 solution spray method for measurement of chloride penetration into hardened cementitious matrix materials. *ACI Materials Journal*. 1992, 89:587–592.

Page CL, Lambert P, Vassie PRW. Investigations of reinforcement corrosion 1: The pore electrolyte phase in chloride contaminated concrete. *Materials and Structures*. 1991, 24:243–252.

Page CL, Vennesland O. Pore solution composition and chloride binding capacity of silica-fume cement pastes. *Materials and Structures*. 1983, 16:19–25.

Pavlík V. Water extraction of chloride, hydroxide and other ions from hardened cement pastes. *Cement and Concrete Research*. 2000, 30:895–906.

Shi C, Yuan Q, Deng D, Zheng K. Test methods for the transport of chloride in concrete. *Journal of the Chinese Ceramic Society*. 2007; 35(4):522–530.

Shi Z, Geiker M R, Lothenbach B, De Weerdt K, Garzon SF, Enemark-Rasmussen K, Skibsted J. Friedel's salt profiles from thermogravimetric analysis and thermodynamic modelling of Portland cement-based mortars exposed to sodium chloride solution. *Cement and Concrete Composites*. 2017, 78:73–83.

Sirivivatnanon V, Khatri R. Chloride penetration resistance of concrete, presented to Concrete Institute of Australia Conf. 'Getting a Lifetime out of Concrete Structures', Brisbane, Australia, October 1998.

Someya K, Daisoku N, Wee T, Nagataki S. Characteristics of binding of chloride ions in hardened cement pastes (in Japanese). *Proceeding of the Japan Concrete Institute*. 1989, 11:603–608.

Stanish K. The migration of chloride ions in concrete. Ph.D. dissertation. Department of Civil Engineering, University of Toronto, Toronto; 2002.

Sugiyama T, Ritthichauy W, Tsuji Y. Simultaneous transport of chloride and calcium ions in hydrated cement systems. *Journal of Advanced Concrete Technology*. 2003, 1:127–138.

Suryavanshi AK, Swamy RN. Stability of Friedel's salt in carbonated concrete structural elements. *Cement and Concrete Research*. 1996, 26:729–741.

Tang L. Chloride transport in concrete—Measurement and prediction. Ph.D. dissertation. Department of Building Materials, Chalmers University of Technology, Goteborg, Sweden; 1996.

Tang L. Electrically accelerated methods for determining chloride diffusivity in concrete—Current development. *Magazine of Concrete Research*. 1996b, 48:173–179.

Tang L, Nilsson LO. Discussion of the paper 'AFREM test procedures concerning chlorides in concrete: Extraction and titration methods', by T. Chaussadent and G. Arliguie. *Materials and Structures*. 2001, 34:128.

Tang L, Nilsson LO. Rapid determination of the chloride diffusivity in concrete by applying an electrical field. *ACI Materials Journal*. 1992, 89:49–53.

Thomas MDA, Matthews JD. Performance of PFA concrete in a marine environment—10-year results. *Cement and Concrete Composites*. 2004, 26(1):5–20.

Viallis-Terrisse H, Nonat A, Petit JC. Zeta-potential study of calcium silicate hydrates interacting with alkaline cations. *Journal of Colloid and Interface Science*. 2001, 244:58–65.

Yonezawa T, Ashworth V, Procter RPM. Pore solution composition and chloride effects on the corrosion of steel in concrete. *Corrosion Engineering*. 1988, 44:489–499.

Yuan Q. Fundamental studies on test methods for transport of chloride ions in cementitious materials. Ph.D. dissertation. Department of Structural Engineering, Ghent University, Ghent, Belgium; 2009.

Yuan Q, Shi C, He F, et al. Effect of hydroxyl ions on chloride penetration depth measurement using the colorimetric method. *Cement and Concrete Research*. 2008, 38:1177–1180.

Chapter 7

Factors Affecting Chlorides Transport in Cement-Based Materials

7.1 INTRODUCTION

When penetrating into cement-based materials, chloride ions will confront both internal environments, such as discrepancies in pore structure and pore solution composition and saturation, and external environments, such as electric field and pressure gradient. The internal and external environments constitute five factors that impact the transport of chloride ions: (1) diffusion, for which the ion gradient depends on single-ion convection; (2) convection, i.e., dissolved ion convection, which is a result of capillary suction or pore solution migration due to pressure gradient; (3) ion binding and adsorption on the pore wall; (4) ion migration due to external electric field accompanied by rebar corrosion; (5) ion movement that is induced by the instant influence of other ions in the pore solution.

In the light of the aforementioned factors, the evaluation of the transport of chloride ions in cement-based materials will be excessively complex. In most scenarios, the study of ion transport in saturated concrete would neglect the influence of convection. Ion types, ion concentrations, pore structure, interfacial transition zone (ITZ), and chloride binding are the factors considered in the analysis of saturated concrete. Meanwhile, cracking is another key factor influencing chloride transport; initiation of cracking in concrete is inevitable when steel-reinforced concrete structures are under the action of mechanical loadings, seasoning, and other physical or chemical attacks. In addition, cracking within a specific width range is commonly allowed in steel-reinforced concrete structures, according to current standards. However, the existence of cracking in concrete would aggravate the ingress of chloride ions, accelerating the deterioration of the reinforced concrete structures. In this regard, the current chapter will discuss the influence of ion types, pore structure, ITZ, chloride binding, and cracking on the transport of chloride in cement-based materials.

7.2 EFFECT OF INTERACTION BETWEEN IONS ON CHLORIDE TRANSPORT

when only chloride migration is considered, Fick law based on the conservation of mass can be used to describe the diffusion of chloride ions in saturation concrete. However, it is worth mentioning that results obtained from Fick law is only a satisfactory approximate of the diffusion in a non-interaction system. In reality, the interaction between ions always exists even in solutions with a low ion concentration, which reduces the chemical energy potential, further decreasing the driving power of the media (Zhang and Gjarv 1995).

Pore solutions of concrete contain various types of ions, such as OH^-, SO_4^{2-}, Na_2^+, K^+, Ca^{2+}. The concentration rank of ions from high to low follows a sequence of OH^-, K^+ and Na^+. According to the ion clouds or Debye–Hückel theory, interaction between ions always exists in water-based solutions because of the properties and structural characteristics of electrolytes (Debye and Hückel 1921; Atkins 1994). The higher the ion concentration, the greater the interaction. Intensive interaction of the ions would be able to significantly impact the corresponding chemical energy potential, when the concentration exceeds a specific value. Therefore, the interaction between ions would decrease the diffusion rate of chloride ions when concrete is exposed to chloride solutions with high chloride concentrations.

7.2.1 Model Describing Multi-Ion Transport

Tang and Nilsson considered the diffusion of chloride ion to follow the Nernst–Planck equation (Tang and Nilsson 1992). The model built by Tang and Nilsson assumes a constant electric field without considering the effects of other ions, which overlooks the influence of chemical activities on the chloride transport due to the simplification. Thus, this model is also called single-ion model (Spiesz et al. 2012).

The ion system in migration experiments is complicated because of the inclusion of multi-ions. A forced application of the conventional model on all the ions would result in an un-neutral charge system (Samson et al. 2003). Therefore, the application of conventional model should be limited to some specific scenarios. To the contrary, the Poisson–Nernst–Planck (PNP) equation is feasible to multi-ion transport with considerations of single conservation of mass and bulk neutral charge. Samson et al. (2003) implemented the following equation to describe the PNP model:

$$j_i^{\pm} = -D_i^{\pm}\left[\frac{\partial c_i^{\pm}}{\partial x} \pm z_i^{\pm}\frac{F}{RT}c_i^{\pm}\frac{\partial \varphi}{\partial x} + c_i^{\pm}\frac{\partial(\ln\gamma_i)}{\partial x}\right] \tag{7.1}$$

$$\frac{\partial c_i^{\pm}}{\partial t} = \frac{\partial j_i^{\pm}}{\partial x} \tag{7.2}$$

$$\varepsilon \frac{d^2\varphi}{dx^2} + F\left[\sum_{i=1}^{n}\left(|Z_i^+|c^+ - |Z_i^-|c^- + \rho\right)\right] = 0 \qquad (7.3)$$

Herein, c_i^+ and c_i^- denote the concentrations of cations and anions, respectively. D_i^- denotes the diffusion coefficient of the ions. ε denotes the absolute dielectric coefficient. ρ denotes the fixed charge density. φ denotes the potential of the electric field.

$\partial\ln\gamma / \partial\ln c$ and $\partial\varphi / \partial\ln x$ are usually neglected in the literature. When an external electric field is applied, Equation 7.1 can be derived as Equation 7.4:

$$j_i^\pm = -D_i^\pm\left[\frac{\partial c_i^\pm}{\partial x} \pm Z_i^\pm \frac{F}{RT} c_i^\pm \frac{\partial\varphi}{\partial x}\right] \qquad (7.4)$$

The potential of the electric field, φ, can be expressed as a combination of external electric field potential and internal potential between ions. The internal potential between ions can be neglected in the case of sufficient external electric field potentials applied. Commonly, other equations, i.e., non-current equation and neutral charge equation, need to be integrated to solve this issue (Lorente et al. 2003). This is because the migration rate of ions is independent with the ion concentration. The calculation of the corresponding migration rate only requires a linear differential equation, which can be solved easily. However, the assumption of neutral charge regards the transport of different types of ions to be mutually independent, which is actually incorrect (Johannesson et al. 2007; Liu 2014). Thus, the Poisson's equation should be implemented rather than the neutral charge equation (Johannesson et al. 2007; Xia and Li 2013; Liu et al. 2012). The Poisson's equation not only considers the non-linear ion transport but also considers the interactions between ions. Therefore, the migration rate of a single ion is time- and position-independent (Liu et al. 2015).

If the static electric potential φ in Equation 7.1 is confirmed according to the external electric field without considering the internal equilibrium of ions charges, i.e., $\partial^2\varphi = 0$, the concentration of each type ion can be calculated based on Equation 7.5. Otherwise, the static electric potential has to be confirmed based on the Poisson's equation (Xia and Li 2013).

$$\partial^2\varphi = -\frac{F}{\varepsilon_0\varepsilon_r}\sum_{i=1}^{N}Z_iC_i \qquad (7.5)$$

Herein, ε_0 denotes the dielectric coefficient in vacuum. ε_r denotes the relative dielectric coefficient at specific temperature. N denotes the sum of all the matters in the solution. It is worth noting that the Equation 7.1 can be approximately replaced by the neutral charge condition, i.e., $\sum_{i=1}^{N} Z_i C_i \approx 0$ (Snyder and Marchand 2001; Lorente et al. 2003 ; Khitab et al. 2005 ; Elakneswaran et al. 2010; Yu and Page 1996; Li and Page 1998; Li and Page 2000; Truc et al. 2000a, 2000b; Wang et al. 2001; Frizon et al. 2003; Toumi et al. 2007; Kubo et al. 2007; Narsillo et al. 2007; Krabbenhoft and Krabbenhoft 2008; Friedmann et al. 2008.). The approximation is based on the great magnitude of 10^{14}(mV mol^{-1}) on the right side of Equation 7.4. $\sum_{i=1}^{N} Z_i C_i \approx 0$ has to be very small, since the value of $\partial^2 \varphi = 0$ is limited. But $\sum_{i=1}^{N} Z_i C_i \approx 0$ does not necessarily lead to $\partial^2 \varphi = 0$, which disproves the validation of the assumption of $\partial^2 \varphi = 0$ based on the neutral charge condition (Xia and Li 2013).

7.2.2 Theory on Interaction between Ions

Equation 7.1 demonstrates that the diffusion coefficient is also determined by the activity coefficient of the diffusion media. The PNP model is more feasible than the conventional single-ion transport model, although it still greatly differs from the practical scenario. According to the Debye–Huckel–Onsager electrolyte theory (Jiang et al. 2013), the conductivity or mobility is also influenced by factors such as chemical activity, cataphoresis, relaxation, etc. (Zhang and Gjarv 1995).

The diffusion of chloride ion in concrete is influenced by various factors, including the characteristics of concrete, the composition of external salt solutions, etc. In the experimental salt solutions, the interactions between ions decreases the chemical potential, which further reduces the driving force of chloride ion diffusion (Zhang and Gjarv 1996). Moreover, other effects in the diffusion system, such as the relatively slower migration rate of cations than anions, the electrical double layer (EDL) on the pore surface, and the pore volume and pore size of the material, can largely impact the chloride ions diffusion (Zhang and Gjarv 1996).

7.2.2.1 The Chemical Potential between Ions

In the past decades, many semi-empirical equations have been developed to compute the chemical activity coefficient of ions in the solution with condensed electrolytes. In these theories, the model proposed by Pitzer is considered to be the most feasible to compute the γ value of condensed solutions (Marchand et al. 1995; Hidalgo et al. 1997). However, the model, as stated by Marchand et al. (1995) and Hidalgo et al. (1997), not only is complicated but also requires important experimental results for the computation. Moreover, it is difficult to solve $\partial \ln \gamma / \partial \ln c$ from the complicated model. The most famed model among these is the Debye–Hückel model (Justnes and Rodum 1997; Bockris and Reddy 1970; Pankow 1994). The main character of the Debye–Hückel model is considering the ions to be infinitely small (assumption of point charges) (Tang 1999a,b). An expanded Debye–Hückel could calibrate the previous models by taking the radius of different ions into consideration (Tang 1999b). The value of $\partial \ln \gamma / \partial \ln c$ can be calculated based on Equations 7.6 and 7.7.

$$\log \gamma = -A \left| z_+ z_- \right| \frac{\sqrt{I}}{1 + B \cdot a^2 \cdot \sqrt{I}} + B^* \cdot I \tag{7.6}$$

$$I = \frac{1}{2} \cdot \sum_i z_i^2 c_i = \frac{n_+ z_+^2 + n_- z_-^2}{2} \cdot c \tag{7.7}$$

Herein, A and B are coefficients related to temperature. A denotes the radii of an ion. B^* is an empirical coefficient dependent upon the type of the solution. I denotes the total charge of the ion, as shown by Equation (7.7).

As shown in Figure 7.1, the activity coefficient decreases with the increase of chloride concentration in the low concentration range (0–1 mol/dm^3) and increases with the increase of chloride concentration in the high concentration range (>1mol/dm^3). As a result, the differential term of $\partial \ln \gamma / \partial \ln c$ experiences a +/− change at a specific chloride concentration, which means that the activity coefficient reduces the effective diffusion coefficient in the

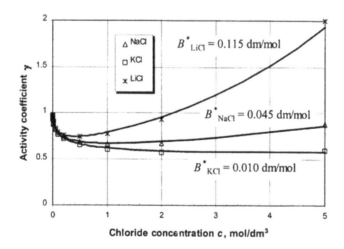

Figure 7.1 Quantitative correlation between γ and c. (From Little 1996.)

low chloride concentration range and increases the effective diffusion coefficient in the high chloride concentration range (Tang 1999b).

7.2.2.2 Lagging Motion of the Cations

During the process of ion migration, the difference in migration rates between cations and anions determines the different migration velocity of ions in electrolyte solutions. Hence, taking the diffusion coefficient of a single ion as the comprehensive diffusion coefficient under both of the influences of cations and anions is not objective, since electrostatic interactions occur between cations and anions. The diffusion of chloride ions is accompanied by the diffusion of cations from the same solution, while the migration velocity of cations is usually slower than the migration of chloride ions. The lagging motion of cations, consequently, would provide a traction backwards for the chloride ions, which can be evidently observed in experiments. This explains the diffusion coefficient of salt in a binary electrolyte solution is a function of both of the diffusion coefficients of cations and anions. In the tests for the chloride ion diffusion coefficient in concrete, the impact of salt type and cation type on the diffusion of chloride ions can be commonly observed (Gjarv and Vennesland 1987; Ushiyama and Goto 1974; Ushiyama et al. 1976; Goto et al. 1979). For instance, at the same level of chloride concentration, the diffusion coefficient of chloride ions rises to 2–3 times of the initial value, when cation in the external salt solution is changed from Na^+ to Ca^{2+}, which can be explained by the higher backwards traction from Na^+ than the traction from Ca^{2+}.

The movement of ions follows the gradient of chemical potential, agreeing with the phenomenon found by Ushiyama and Goto (1974) and Roy et al. (1986) that the diffusion rate of cations (e.g., Na^+, K^+, Li^+, Cs^+, etc.) is inferior to that of chloride ions. The neutral charge of the solution determines that the chloride ions is always surrounded by cations. Once the chloride ions move forward, an anti-electric field forms between the chloride ions and the lagging surrounding cations. The anti-electric field will lead to a redistribution consequently (Tang 1996). The driving force (chemical potential) and the traction (anti-electric field), together, are called "electrochemical potential," which explains the movement of ions under the gradient of electrochemical potential.

Tang (1999a) deemed that the diffusion of chloride ions in concrete results in an effect similar to an ion-selective semipermeable membrane, which introduces an extra anti-electric field. Therefore, the anti-electric field in the solution–concrete system consists of two parts: One part comes from the external solution K_{t0}, and another part comes from concrete K_{tm} (Zhang and Buenfeld 1997). Zhang and Gjorv (1995) brought out that the interaction between ions generates two effects, i.e., cataphoresis and relaxation, under external electric fields, which are the cause of tractions.

7.2.2.3 Interaction between the EDL and Ionic Clouds

In the solution system with solid electrolyte, the formation of EDL on the solid surface is also an important phenomenon. The model of EDL is presented in Figure 3.12 in Chapter 3. According the model, the EDL on the surface of capillary pores would disturb the ion clouds in the solution, which further impact the movement of ions and fluids. Therefore, the ion diffusion in the porous cement-based material would also be influenced by the effect of EDL. Relative to the Debye length, the finer the capillary pores, the greater the effect of EDL; this results in a slower migration rate. In addition, an increase in Debye length would lead to a decrease in ion migration rate in the same capillary pore (Zhang and Gjorv 1996).

On the surface of capillary pores, the fluid is static due to the effect of EDL. Meanwhile, a repulsion potential barrier forms because of the EDL on the charged solid surface. For a given solid, the repulsion potential barrier is mainly determined by the property of the electrolyte solution and the thickness of the EDL, since the van der Waal's force is constant (Zhang and Gjorv 1996). When the charges of the ion clouds and the EDL on the capillary surface is the same, the repulsion potential barrier would inhibit the chloride ions from penetrating into the capillary pores. Therefore, for capillary pores with a given dimension, a minimum concentration is likely required for the diffusion of ions into concrete (Zhang and Gjorv 1996). For a given ion concentration, chloride ions are likely only able to penetrate into capillary pores with a size greater than a specific value. If the size of capillary pores turns to low enough, chloride ions have a low likelihood to diffuse into concrete because of the incompressibility of the EDL and ion clouds.

The application of silica fume and GGBFS could effectively diminish the pore size of concrete, which promotes the resistance of concrete against chloride diffusion (Gjorv and Vennesland 1979; Frey et al. 1994). This could be one of the reasons why the chloride diffusion in high-performance concrete predicted by the Fick's law significantly differs from the experimentally observed results (Genin 1986). However, when the vast majority of the pores possess a dimension sufficiently greater than the ion radius, the quantity of ions passing through unit area within unit time would roughly positively correlate to the pore volume, but negatively correlate to the square of the ion cloud thickness (Zhang and Gjorv 1996).

7.2.3 Concentration Dependence of Chloride Transport

Many researchers (MacDonald and Northwood 1995; Bigas 1994; Arsenault et al. 1995; Achari et al. 1995; Zhang 1997; Zhang and Gjørv 1996) utilized solutions with different concentration of chlorides to prove

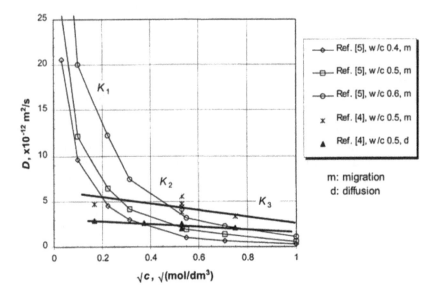

Figure 7.2 Non-linear correlation between the diffusion coefficient and square root of chloride concentration. (Tang 1999b.)

the dependence of chloride diffusion in concrete to the chloride concentration. Chatterji (1994) proposed a rule of "square root" to describe this dependence to the chloride concentration, as shown in Equation (7.8).

$$D = D_0\left(1 - K\sqrt{c}\right) \qquad (7.8)$$

However, the rule of "square root" is not always applicable in predicting the chloride diffusion in concrete. As shown in Figure 7.2, the slope of the curve (K value) changes with the chloride concentration of the external solution, which means that K is not a constant but related to the ion concentration. Therefore, it can be concluded that Equation (7.8) is only applicable in a very narrow range of chloride ion concentration. Some researchers (Andrade 1993; Marchand et al. 1995) attempted to interpret the dependence of the diffusion rate to ion concentration by the activity coefficient; however, the activity coefficient only varied in a range of 1 to 0.68, when the NaCl concentration varied from extremely low to 1 mol/dm^3 (Andrade 1993). It is difficult to explain the discrepancy in the diffusion coefficient of chloride ions in Figure 7.2 solely by the activity coefficient, but Jiang et al. (2013) found that the influence of solution concentration on the non-steady-state ion diffusion is insignificant.

7.2.4 Effect of Species on Chloride Transport

Ushiyama and Goto (1974) investigated the influence of material type on the cement paste, while the influence of concentration is not directly compared. Difficulties for direct comparison are ascribed to the complexity and non-stability of the cementitious system as well as the impediment in controlling the solution type in pores. The degree of reaction can only be accurately estimated based on the determination of the activity of each matter (Reardon 1990; Duchesne and Reardon 1995). Additionally, an accurate model of combination and chemical reaction not only requires the determination of the activity of ions but also has to consider the time-dependence among transport, adsorption, and chemical reaction (Barbarulo et al. 2000). The accuracy of the model would be discounted, if the detailed diffusion information is unavailable but only the matter activity can be accurately estimated.

Snydera and Marchand (2001) employed the electro-diffusion equation as the model to compute the ion transport. They found that the matter type has a relatively more significant impact on the apparent diffusion coefficient of ions, compared to concentration, in non-active porous materials. Specifically, in the system studied by Snydera and Marchand (2001), the concentration altered the apparent diffusion coefficient within $\pm 20\%$. In comparison, the matter type was able to increase the apparent diffusion coefficient of non-active system to twice the initial value in a relatively short time. Some matters even enabled the apparent diffusion coefficient to exhibit time-dependence within a long amount of time. These aforementioned systems are not suitable to be modeled based on the Fick's law and constant apparent diffusion coefficient. For the system studied, the apparent diffusion coefficient diverges to infinite due to the potential diffusion and would eventually turn to negative values.

According to these mechanisms, such as the influence of reduced chemical potential, delayed migration rate, and the double electrical layer, Zhang and Gjorv (1996) proposed a method to estimate the relative diffusion rate of chlorides in different types of salt solutions. Equation 7.9 tells a sequence rank of diffusion rates for four classic salt solutions.

$$D_{LiCl} < D_{NaCl} < D_{Kcl} < D_{CaCl_2} \tag{7.9}$$

The sequence in Equation 7.9 complies with most experimental results reported in literature (Gjarv and Vennesland 1987; Ushiyama and Goto 1974; Ushiyama et al. 1976). But the sequence is estimated based on solutions with much lower concentrations than the solutions commonly used to test the chloride diffusion coefficient in concrete. The ion clouds in the solution would lose their smoothness at higher concentrations, which invalids the applicability of the Debye–Hückel theory. A "quasi-crystal" theory was introduced regarding this scenario (Bockris and Reddy 1977), but its application is limited for application so far.

7.3 EFFECT OF MICRO-STRUCTURE ON CHLORIDE TRANSPORT

7.3.1 Effect of Pore Structure on Chloride Transport

The chloride diffusion coefficient in concrete and cement paste varies in a wide range, which largely depends on the water/cement (w/c) ratio of the cement-based material (Pivonka et al. 2004). It has been widely accepted that the cement-based materials with different w/c ratios possess different pore structure. This implies that the pore structure would significantly impact the chloride diffusion coefficient of porous cement-based materials (Mohammed et al. 2014). Some researchers (Yang 2006; Yang et al. 2006; Sun et al. 2011) investigated the influence of pore structure on the transport of chloride ions. Parameters of the pore structure of cement-based materials, which impacts the transport of chloride ions, include average pore diameter, threshold pore diameter, interconnected pore diameter, capillary pore volume, capillary porosity, bulk porosity, tortuosity, blocking rate, and so on. Maekawa et al. (2003) regarded the dimension of interlayer pores in molecular magnitude, which is much smaller than the diameter of ions; therefore, they consider no ion communication in interlayer pores of cement paste.

7.3.1.1 Relation between Porosity and Chloride Transport

Figure 7.3 presents the relationship between the chloride diffusion coefficient and porosity. The figure indicates that the chloride diffusion coefficient significantly increases when the porosity exceeds 0.18. This phenomenon may be a result of the discrepancy in pore size distribution, porosity, and specific surface area ascribed to the different components in concrete. A further decrease in porosity and pore diameter is expected to continuously reduce the chloride diffusion coefficient, even to a significantly low value. Pore structures can be categorized into three regions (Li et al. 2018), according to Equation 7.10, shown as in Figure 7.3 (right). Region I—subcritical state: $\varepsilon \leq \varepsilon_{cr}$, i.e., the porosity is lower than the percolation threshold. In this circumstance, pores cannot form an interconnected network, which are usually regarded as isolated pores and lock pore networks. Region II—critical state—around the Region I—subcritical state: $\varepsilon_c \leq \varepsilon \leq 1.5\varepsilon_{cr}$. Region III—conventional migration region: $\varepsilon \geq 1.5\varepsilon_c$, i.e., the porosity of concrete is far greater than the percolation threshold, where all the pores are involved in the migration process.

$$D_f = D_0 (\varepsilon - \varepsilon_{cr})^n \, n > 1 \tag{7.10}$$

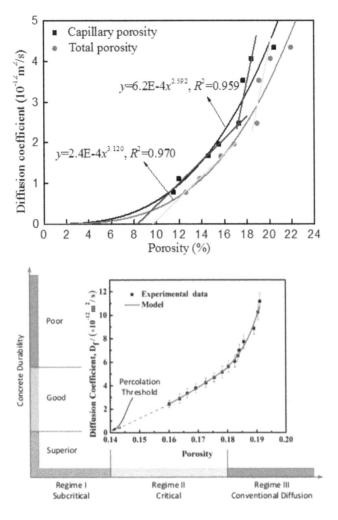

Figure 7.3 The influence of concrete porosity on the chloride diffusion coefficient. Top part refers to Yang (2006), and the bottom part refers to Li et al. (2018).

where D_f denotes the characteristic diffusion coefficient (m²/s); ε and ε_{cr}, respectively, denote the porosity and the critical porosity. n is the exponential value.

However, it can be seen from Figure 7.4 that Region I seems not to exist. This elucidates that the chloride diffusion coefficient cannot decline to 0 when the capillary porosity is infinitely small, which may be ascribed to the chloride migration in a relatively lower rate through the pores finer than capillary pores. Therefore, it is actually possible to prohibit or inhibit the corrosion to a large extent (e.g., reduce two magnitudes), if concrete with a porosity lower than the critical value can be designed. However,

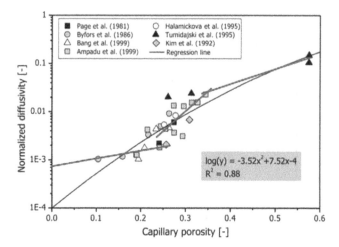

Figure 7.4 The relationship between capillary porosity of cement paste and regressed diffusion coefficient. (From Choi et al. 2017.) (Red lines are added by the authors.)

this subcritical state is difficult to realize, since the excessive water needs open routs to be released from the matrix, which leads to inevitable porosities. Moreover, the tortuosity significantly declines as the porosity increases (Sun et al. 2011), which possibly explains the evidently low chloride diffusion coefficient obtained from a relatively low porosity.

7.3.1.2 Relation between Pore Diameter and Chloride Transport

The relationship between the pore diameter and the chloride diffusion coefficient has a similar trend with the relationship between the porosity and the diffusion coefficient, as shown in Figure 7.5. Li et al. (2018) also proposed three regions where the diffusion coefficient varies as the pore diameter increases, which agrees with the three categorized porosity regions. As can be seen from Figure 7.5, the chloride diffusion coefficient rapidly increases when the pore diameter exceeds 100 nm, which may be due to the rapid increase of blocking ratio beyond some specific pore diameters (Sun et al. 2011a). The major migration path in hardened cement paste is capillary pores, while the gel pores only plays a secondary/minor role unless the capillary pore volume is extremely low (Garboczi and Bentz 1992). Therefore, it is necessary to consider the evident influence of blocking rate in the study of effect of pore structure on the chloride diffusion coefficient. When taking tortuosity and blocking into consideration, Sun et al. (2011) obtained similar diffusion coefficients from their model to the experimental data. It was demonstrated that the evaluation of the pore structure's effect on the

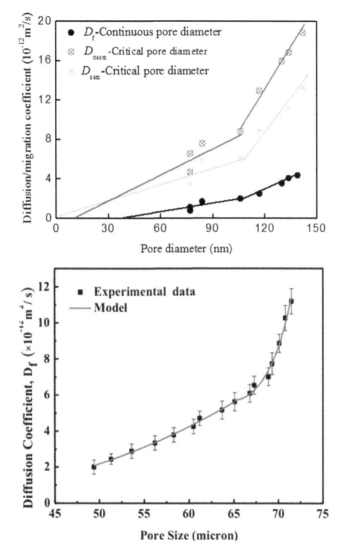

Figure 7.5 Relationship between pore diameter and chloride diffusion coefficient. Top part refers to Yang (2006), and the bottom part refers to Li et al. (2018).

chloride diffusion coefficient would be more realistic with the integration of effect of tortuosity and blocking.

7.3.2 Effect of ITZ on Chloride Transport

In conventional concrete, the w/c gradient of the interfacial layer causes the different microstructure of matrix surround aggregates (Aitcin and Mehta 1990).

This hydration product matrix surround aggregate is called ITZ, which was first found by Farran (1956). ITZ contains a higher concentration of crystalline $Ca(OH)_2$ and a higher porosity around aggregates, compared to the further matrix from aggregates, which results in a higher chloride diffusion coefficient through ITZ than the chloride diffusion coefficient through cement paste.

7.3.2.1 Transport Properties of Chloride in ITZ

Figure 7.6 illustrates the simulated distribution of unhydrated cement grains, capillary pores, gel pores, and hydration production in concrete with a w/c = 0.4 after 28 days of hydration. It can be seen from Figure 7.6 that both the porosity and capillary porosity increase with decreasing x (distance from aggregate), when x <20 μm. Based on the discussion in Section 7.3.1, the increase of porosity and capillary porosity leads to the increase of chloride diffusion coefficient, which means a decrease of chloride diffusion coefficient with increasing x, shown as in Figure 7.7.

Tian et al. (2018) investigated the ITZ between cement paste and cylindrical aggregate by electromagnetic pulse (EMP), identifying an approximate ITZ thickness of 40.5 μm that decreases along two sides. Statistic analysis demonstrates that the thickness of ITZ follows normal distribution, yielding an average value of 40.7 μm from all the tested results. Meanwhile, the study finds the ITZ thickness of cylindrical aggregates with a diameter of 5 mm, 7 mm, and 10 mm to be 40.9 μm, 40.6 μm, and 41.1 μm, respectively. No evident difference in ITZ thickness is noticed for aggregates with different sizes, which agrees with Scrivener et al.'s results (2004). This is possibly because the thickness of ITZ is mainly determined by the median size of the cement grain rather than the size of the aggregate (Scrivener et al. 2004).

Actually, different ITZ thicknesses are measured by different researchers, which is determined by the experimental techniques and analytic models. Basheer and Kropp gave a general evaluation on the thickness of ITZ in a range of 0–100 μm (Basheer and Kropp 2001). The thickness range was further narrowed into 40–50 μm by the analysis based on SEM (Bentz et al. 1992). Evidently, the test results of microprobe (Tian et al. 2018) agreed with the results of SEM (Bentz et al.1992). Nevertheless, most studies still take the range of 10–50 μm as the representative thickness of ITZ. This thickness range maybe lead to significant difference among effects of ITZ on chloride transport, demonstrated by Figure 7.7.

7.3.2.2 Effect of Aggregate Volume on Chloride Transport

Theoretically, the volume of ITZ should have significant influence on the migration of chloride ions, which has also been proven by many researchers by varying aggregate volume. The chloride diffusion coefficient is found to

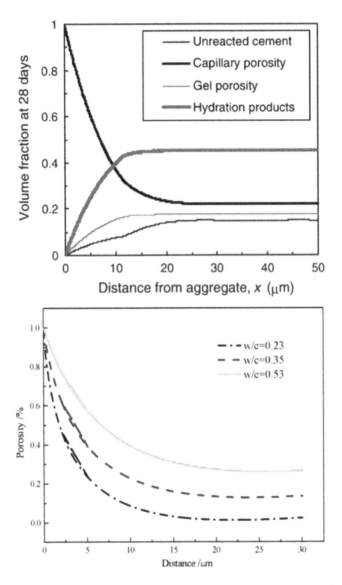

Figure 7.6 Top: Schematic of simulated gradient in concrete after 28 days of hydration (w/c=0.40, ASTM I cement, aggregate size: 0.15–16mm, h=25μm) (Zheng et al. 2009). Bottom: Simulated distribution of ITZ porosity. (From Sun et al. 2011b.)

continuously decrease when the volume of aggregates keeps increasing, as shown by the results in the literature (Yang and Su 2002; Delagrave et al. 1997; Zheng et al. 2012; Zheng and Zhou 2013; Zheng et al. 2018; Wang et al. 2018). The increase of aggregate volume signifies the increase of ITZ volume. Thus, the ITZ volume has an evident impact on the migration of

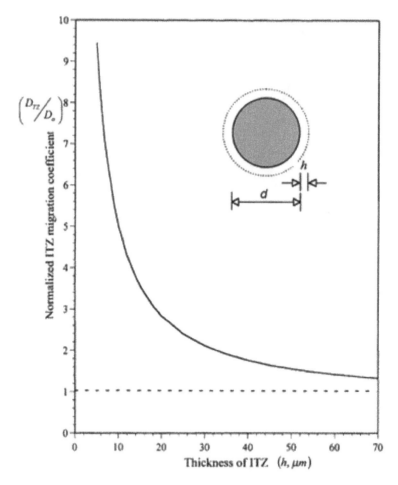

Figure 7.7 Relationship between ITZ thickness and ITZ chloride diffusion coefficient regressed from the overall chloride diffusion coefficient of the matrix. (From Yang and Su 2002.)

chloride ions. In other words, except for the thickness of ITZ, the surface area of ITZ also has an evident impact on the migration of chloride ions.

As mentioned above, the thickness of ITZ mainly varies in the range of 10–50 μm, and ITZ possesses a different chloride diffusion coefficient than the one of neighbouring cement matrix. Moreover, the chloride diffusion through aggregate is regarded negligible because of the densified microstructure of aggregate. As a result, even though the addition of aggregates introduces more ITZ, which may increase the chloride diffusion coefficient, the dilution and distortion effect of adding aggregates would compensate for the negative effect of ITZ on increasing the chloride diffusion

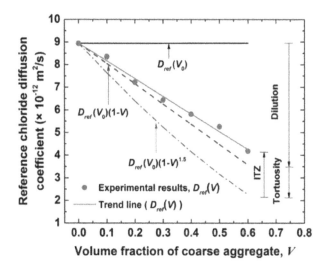

Figure 7.8 Relationship between chloride diffusion coefficient and volume fraction of coarse aggregate. (From Wang et al. 2018.)

coefficient and eventually decrease the bulk the chloride diffusion coefficient of concrete (Wang et al. 2018), as shown in Figure 7.8.

7.3.2.3 Effect of Aggregate Shape on Chloride Transport

The shape of aggregate affects the property of ITZ, further influencing the chloride diffusion coefficient, which can be approved by Figure 7.9. It can be seen from Figure 7.9 that D_{con}/D_{cp} increases as D_{itz}/D_{cp} increases for elliptical aggregates with a constant aspect ratio (u). For a given D_{itz}/D_{cp}, D_{con}/D_{cp} decreases as u increases, which is mainly due to the fact that the surface area fraction of ITZ is a monotonic decreasing function of u (Zheng et al. 2012). It can be seen from Figure 7.9 that concrete made from aggregates with an identical size has the maximal chloride diffusion coefficient. This denotes that the adjustment of aggregate shape would modify the surface area of ITZ, which would further adjust the chloride diffusion coefficient of concrete. The evaluation of aggregates' influence on chloride diffusion coefficient needs to consider aggregate volume fraction and shape together with aggregate gradation.

7.3.3 Coupling Effect of Pore Structure and ITZ on Chloride Transport

It is illustrated in Figure 7.10 that the effective chloride diffusion coefficient is a function of capillary porosity in concrete with different aggregate volume fractions V_a. If the aggregate shape is assumed to be spherical, the

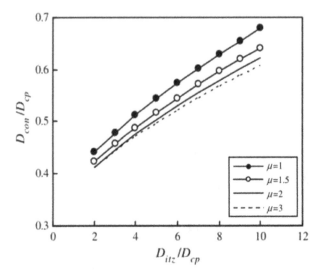

Figure 7.9 Influence of aggregate shape on chloride diffusion coefficient. (From Zheng et al. 2012.)

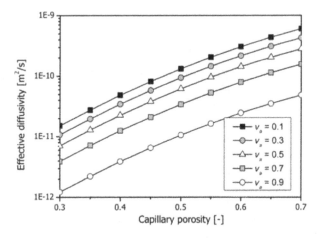

Figure 7.10 Relationship between chloride diffusion coefficient and capillary porosity in concrete. (From Choi et al. 2017.)

development of effective chloride diffusion coefficient exhibits an evident increase as the capillary porosity increases. Figure 7.10 also demonstrates the decrease of chloride diffusion coefficient with increasing the aggregate volume fraction. Because the migration of chloride ions in aggregates is negligible, the increase of aggregation content in concrete reduces combined chloride diffusion coefficient of ITZ and matrix due the dilution and distortion effect of adding aggregates. Figure 7.11 presents that the

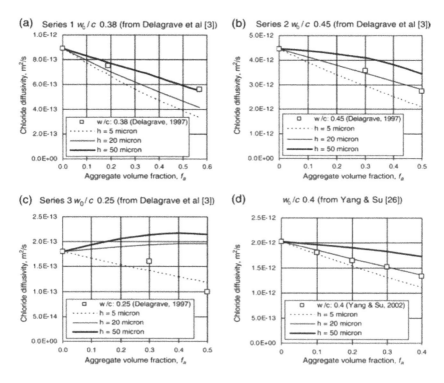

Figure 7.11 Comparison of simulation analysis and experimental results in Delagrave et al. (1997) and Yang and Su's (2002) studies. (From Zheng et al. 2009.)

chloride diffusion coefficient of concrete with w/c ratios of 0.38, 0.40, and 0.45 decreases when the aggregate volume increases, while the chloride diffusion coefficient of concrete with a w/c ratio of 0.25 increases with the aggregate volume increasing. It is demonstrated that the dilution and distortion effect of adding aggregates does not evidently decrease the chloride diffusion coefficient of concrete with a low porosity. Although there is no data demonstrating the coupling effect of ITZ and pore structure, it is worth noting that the addition of aggregates would introduce new connected path between ITZ and the pores in cement paste; therefore, the coupling effect of ITZ and pore structure of cement paste should not be neglected in the chloride diffusion resistance design of concrete.

7.4 EFFECT OF CHLORIDE BINDING ON CHLORIDE TRANSPORT

In most field cases, chloride ion may penetrate into concrete through a combined mechanism of hydraulic advection, capillary suction, diffusion, and thermal migration. Diffusion is the dominating mechanism in the case of saturated concrete, such as concrete submerged in seawater. In the past decades, many models (Olivier 2000; Tang 1996; Xi and Bažant 1999), based on diffusion alone were developed to predict chloride penetration into saturated concrete. Also, some models (Ababneh et al. 2003; Marchand 2001; Saetta et al. 1993) based on combined mechanisms were developed to predict chloride penetration into unsaturated concrete. This section focuses on the diffusion of chloride ion in saturated concrete. As discussed in the introduction, chloride binding has a significant effect on the chloride transport process. It retards chloride ion penetrating into concrete and prolongs the time for corrosion initiation. Therefore, in order to better predict the chloride transport process, chloride binding should be taken into account in the models.

7.4.1 Describing Effect of Chloride Binding on Chloride Transport by Binding Isotherm

When chloride ion transports through diffusion alone, the chemical gradient is the only driving force. The chemical potential of chloride is given by (Philip 1994, Zhang and Gjørv 1996):

$$\mu = \mu_0 + RT\ln(\gamma c) \tag{7.11}$$

The movement of ionic species is under the gradient of electrochemical potential, which is the combination of driving force (chemical potential) and draw-back force (counter electrical field) (Tang 1999), can be written as

$$J = -\frac{D}{RT}c\nabla\mu = -D\frac{\partial c}{\partial x}\left(1 + \frac{\partial \ln y}{\partial \ln c}\right) - cD\frac{zF}{RT}\frac{\partial \Phi}{\partial x} \tag{7.12}$$

For the purpose of simplification, in the literature, the terms $(1 + \frac{\partial \ln y}{\partial \ln c})$ and $cD\frac{zF}{RT}\frac{\partial \Phi}{\partial x}$ are often neglected. Equation 7.12 becomes Fick's first law:

$$J = -D\frac{\partial c}{\partial x} \tag{7.13}$$

For a non-steady-state process, one gets Fick's second law (Crank 1975),

$$\frac{\partial}{\partial x}\left(D\frac{\partial c}{\partial x}\right) = -\frac{\partial J}{\partial x} = \frac{\partial c_t}{\partial t} = \frac{\partial c_f}{\partial t} + \frac{\partial c_b}{\partial t} = \frac{\partial c_f}{\partial t}\left(1 + \frac{\partial c_b}{\partial t}\right) \tag{7.14}$$

Fick's second law can be used to describe the chloride transport. In cases when chloride binding is present, the bound chloride ion will be removed from the diffusion flux and can be subtracted from the conservation of mass equation:

$$\omega_e \frac{\partial c_f}{\partial t} = \frac{\partial}{\partial x} D_e \omega_e \frac{\partial c_f}{\partial x} - \frac{\partial c_b}{\partial t} \tag{7.15}$$

$$c_t = c_b + c_f \cdot \omega_e \tag{7.16}$$

where c_f is the free chloride concentration (kg/m^3 solution), c_b is the bound chloride concentration (kg/m^3 concrete), c_t is the total chloride concentration (kg/m^3 concrete), and D_e is the effective diffusion coefficient. x_e is the evaporable water content (m^3 solution/m^3 concrete). Inserting Equation 7.16 into 7.15, one obtains,

$$\frac{\partial c_t}{\partial t} = \frac{\partial c_f}{\partial t}\left(\omega_e + \frac{\partial c_b}{\partial c_f}\right) = \frac{\partial}{\partial t}\left(D_e \omega_e \frac{\partial c_f}{\partial x}\right) \tag{7.17}$$

Thus

$$D_a = \frac{D_e}{\left(1 + \frac{1}{\omega_e}\frac{\partial c_b}{\partial c_f}\right)} \tag{7.18}$$

The influence of various binding isotherms on D_a is shown in the following equation:
No binding:

$$C_b = 0, \; D_a = D_e$$

Linear isotherm:

$$C_b = kc_f, \frac{\partial c_b}{\partial c_f} = k, D_a = \frac{D_e}{1 + \frac{1}{\omega_e}k} \tag{7.19}$$

Freundlich isotherm:

$$c_b = \alpha c_f^\beta, \frac{\partial c_b}{\partial c_f} = \alpha\beta c_f^{\beta-1}, D_a = \frac{D_e}{1 + \frac{1}{\omega_e}\alpha\beta c_f^{\beta-1}} \tag{7.20}$$

Langmuir isotherm:

$$c_b = \frac{\alpha c_f}{1 + \beta c_f}, \frac{\partial c_b}{\partial c_f} = \frac{\alpha}{\left(1 + \beta c_f\right)^2}, D_a = \frac{D_e}{1 + \frac{\alpha}{\omega\left(1 + \beta c_f\right)^2}} \tag{7.21}$$

Martın-Pérez et al. (2000) solved the conservation of mass equations by using one-dimensional finite-difference method. The initial and boundary conditions used for the numerical analysis were the following:

$$t = 0, c_f = c_0, \text{ at } x > 0$$

$$t > 0, c_f = c_s, \text{ at } x = 0$$

$$t > 0, c_f = c_0, \text{ at } x = L$$

L is the thickness of the concrete sample, t is the time, c_s is the surface chloride concentration, and c_0 is the background concentration.

Three binding scenarios were considered in the model: (1) no binding; (2) linear binding isotherm; (3) Freundlich binding isotherm. It was assumed that the effective diffusion coefficient D_e was $1.0 \times 10^{-12} m^2/s$, the surface chloride concentration was 0.5 M, and ω_e was 8%.

Figures 7.12 and 7.13 show the predicted free and total chloride concentration profiles after 5 years and 50 years of exposure to 0.5 M chloride solution. As shown in the predicted total chloride concentration profiles, the order of the free chloride concentration at a given depth is: Freundlich binding < linear binding < no binding. This means that when no binding is considered in the model, free chloride concentration reaches the threshold earliest. This may result in an underestimation of the concrete structure's service life. If no binding is accounted for in the model, the total chloride concentration is higher than that when binding is considered in the model. Based on the results from Martın-Pérez's, we can understand how significantly the calculated chloride profiles depend on the assumed binding isotherm used in the model. Subsequently, the binding isotherms affect the predicted penetration depth and the time for corrosion initiation.

The model mentioned above is the simplest one, in which many factors influencing the chloride binding and chloride transport were neglected, such as pH, temperature, age, etc. In fact, the continuous hydration of

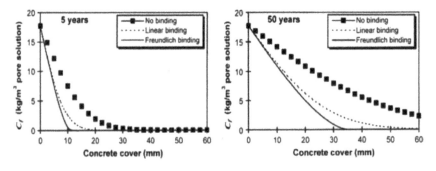

Figure 7.12 Predicted free chloride concentration profiles at 5 years and 50 years for 0.5 M exposure condition. (From Martın-Pérez et al. 2000.)

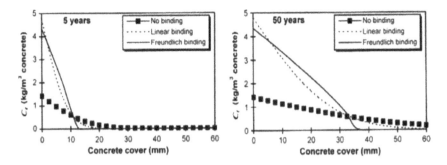

Figure 7.13 Predicted total chloride concentration profiles at 5 years and 50 years for 0.5M exposure condition. (From Martın-Pérez et al. 2000.)

cement affects the chloride binding as well. The change of pH due to leakage of hydroxyl is also known to influence the relationship between free and bound chloride. Because of these simplifications, the model used in Martın-Pérez's paper, whose original purpose is to use a simplified model to clarify the chloride binding on service life prediction, is not appropriate for the service life prediction, Tang (1996) developed a model, called ClinConc, to predict chloride transport in concrete, in which a Freundlich binding isotherm was implemented. The pH and the temperature effects were also considered. The equations were solved using a finite-difference method. Olivier (2000) also developed a more complicated model, called MsDiff, in which the interactions of four types of ions were considered. He applied the Freundlich binding isotherm. Although these models are sophisticated, chloride binding may have similar influences on the predicted results of these models.

Instead of using binding isotherms, Samson and Marchand considered only chemical binding, but neglected physical binding (Marchand 2001, Samson et al. 2000). The chemical equilibrium of various solid phases, including portlandite, ettringite, Friedel's salt, calcium–silicate–hydrate (C–S–H), gypsum, etc, present in cement material is verified at each point by considering the concentrations of all ions. A local chemical equilibrium throughout the system was assumed. These assumptions are questionable, because physical binding may also play an important role in the binding phenomena.

7.4.2 Transportable Chloride in Cement-Based Materials

An overview of the EDL model is presented in previous publications (Friedmann et al. 2008; Bard and Faulkner 2001). When liquid is in contact with a solid surface, an EDL will form. The EDL model can be described by the Helmholtz model, Gouy-Chapman model, Grahame model, and Stern model (a special case of the Grahame model). The Helmholtz model, which describes capacity as a constant, fails to correctly explain some experimental results dealing with the variation of the capacity in the function of the difference of potential between solid phase and liquid phase (Bard and Faulkner 2001; Galus 1994). In the Gouy–Chapman model, ions are regarded as punctual charges that can approach the solid phase until the distance between them and the interface is zero. However, in the Gouy–Chapman model, the calculated capacities are not in agreement with experimental results (Bard and Faulkner 2001; Galus 1994).

In the Stern model, the charged domain is structured as a diffuse layer and a compact layer. Grahame improved the Stern model by considering the possibility of chemical adsorption ("specific adsorption") (Friedmann et al. 2008). In this type of interaction, the charges of the adsorbed ions could be the opposite to that of the EDL. The compact layer is divided into two parts: the inner Helmholtz plane, defined as the center of specifically adsorbed ions, and the outer Helmholtz plane, defined as the center of non-specifically adsorbed ions. Chloride in diffuse layer can be transportable and thus called as transportable chloride (Friedmann et al. 2008; He et al. 2016).

It was reported that Friedel's salt was unstable when the pH was lower than 12 (Birnin-Yauri and Glasser 1998; Suryavanshi and Swamy 1996). In fact, even keeping pH>12 and Friedel's salt stable, it is not reasonable that only free chloride transport was considered due to transportable physically bound chloride in the diffuse layer. It is worth noting that transportable chloride concentration may be more accurate than free chloride concentration to evaluate durability of concrete exposed chloride environment.

7.5 EFFECT OF CRACKING ON CHLORIDE TRANSPORT

7.5.1 Cracking Formation Method in Laboratory Studies

Artificial cracking is usually implemented in laboratory studies by many researchers to overcome the complexity in crack characterization. Cracking that forms in the laboratory can be divided into two categories: artificial cracking (e.g., cracking initialized by pre-set notches) and natural cracking (e.g., cracking initialized by preloading). The major advantage of artificial cracking includes the ease of fabrication, the ease to define the parameters (e.g., crack width) and convenience of establishing analysis models (Marsavina et al. 2009; Mu et al. 2013). But the disadvantage of artificial cracking is also obvious. The smooth surface of artificial cracking differs from the tortuous and coarse cracking in reality. Moreover, the surface of pre-set notches contains more cement paste than natural cracking, and artificial cracking is relatively wider than natural cracking. Methods of generating natural cracking includes: splitting (Wang et al. 1997; Dai et al. 2010) and mechanical inflation (Ismail et al. 2008), by which traction cracking is generated on concrete plate. However, natural cracking generated by these methods have a uniform width (Wang et al. 1997). Another method to generate natural traction crack is based on the three-point or four-point flexural test (Gowripalan et al. 2000; Wittmann et al. 2009; Sahmaran 2007), which produces cracking closer to real cracking. There is yet no standardized method to characterize cracking, which warrants further studies.

7.5.2 Characterization of Cracking

To establish the quantitative correlation between cracking width and chloride diffusion coefficient, which is similar to the correlation between the diffusion in hardened cement paste and pore structure (Garboczi 1990), the chloride diffusion coefficient in one single crack is defined as:

$$D_{cr} = \beta_{cr}D_0 \tag{7.22}$$

where, β_{cr} is defined as "geometrical factor of cracking," which considers the tortuosity, connectivity, and shrinkability of the cracking path perpendicular to the diffusion direction. D_{cr} and D_0 denote the chloride diffusion coefficient (m²/s) in cracking and solutions, respectively. To put it differently, cracking can be regarded as a straight path with a diffusion coefficient $D_{cr} = \beta_{cr}D_0$, as shown in Figure 7.14. Herein, the cracking geometrical factor β_{cr} is the reciprocal of the tortuosity that is suggested by Gérard et al. (1997).

In recent years, it has been revealed that the cracking density, cracking direction, cracking tortuosity, and other cracking parameters, in addition to cracking width and depth, also have vital influence on the chloride diffusion coefficient in concrete (Wang et al. 1997; Ishida et al. 2009; Akhavan et al. 2012; Zhou et al. 2011; Wang et al. 2016a). Even the roughness of cracking surface could also impact the chloride diffusion coefficient (Ye et al. 2012; Rodriguez and Hotoon 2003). However, it is still a challenge to establish a model to take these parameters into consideration and correlate them with the determination of chloride diffusion coefficient in concrete. The cracking width has been reported in the literature to be the key factor that influences the transport properties of concrete. Different from notch cracking with uniform widths, natural cracking generated from mechanical loading are in a random distribution, as shown in Figure 7.15(a) and (b). The width of natural cracking varies along both the surface and propagation direction. In addition to cracking width and depth, cracking direction, cracking tortuosity, and cracking shrinkability, cracking density is also an essential parameter that influences the chloride diffusion coefficient in concrete.

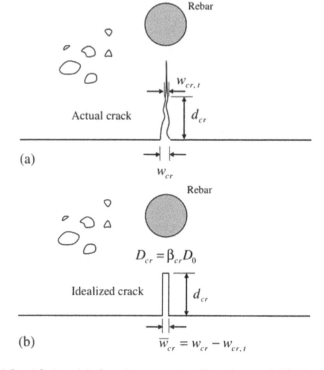

Figure 7.14 Simplified model of crack propagation. (From Jang et al. 2011.)

7.5.3 Effect of Cracking on Chloride Transport

7.5.3.1 Effect of Cracking Width

Aldea et al. (1999) investigated the influence of cracking on the water and chloride ions permeability in concrete. In their study, cracking widths in a range of 50–250 µm was generated by splitting traction experiments, and the chloride ion permeability was determined by the rapid chloride permeability test (RCPT) (Witting 1981). The results indicated that the chloride ions' permeability increased with the cracking width increasing. Rodriguez and Hotoon (2003) studied the influence of cracking width and cracking surface roughness on the chloride ions migration rate. Cracking widths in a range of 80–680 µm is generated by splitting traction experiments, and non-steady-state diffusion test (NT Build 443 1995) was conducted to evaluate the diffusion coefficient. Experimental results indicated that the diffusion of chloride ions is independent, with cracking width and cracking surface roughness; in the case of cracking widths, ranging from 80 to 680 µm. François et al. (2005) investigated the influence of slit cracking that was generated by mechanical inflation on the local diffusion in cement

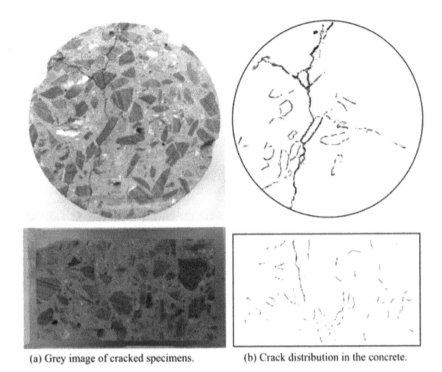

(a) Grey image of cracked specimens.   (b) Crack distribution in the concrete.

Figure 7.15 Digital image analysis in recognizing cracking. (From Wang et al. 2016a.)

paste. Chloride ion concentration in the region perpendicular to the cracking path was determined after 15-day immersion of the specimens in the chloride solution. The results revealed that the chloride diffusion through the cracking wall is similar to the diffusion through specimen surface if the cracking width is greater than 205 μm.

Djerbi et al. (2008) confirmed the relationship between cracking width and chloride diffusion coefficient through cracking. D_{cr} linearly increases as cracking width increases from 30 μm to 80 μm, while D_{cr} nearly stagnates as cracking width exceeds 80 μm. The results indicate that materials, even with different tortuosity and roughness, have no influence on D_{cr} (Rodriguez and Hotoon 2003).

Wang et al. (2016a) suggested that the chloride diffusion depth in the cracking region is determined by cracking width and cracking tortuosity. With a same surface cracking width, the tortuosity of the primary cracking determines the diffusion depth of chloride ions. Therefore, Wang et al. (2016a) defined effective cracking width by adding tortuosity into cracking width so as to correlate cracking width and chloride ion migration rate.

7.5.3.2 Effect of Cracking Depth

Ye et al. (2013) observed a rapid decrease of chloride ion concentration from the surface of cracking to the interior of concrete. At a same distance from the cracking surface, a higher chloride ion concentration can be measured with a higher chloride ion concentration on the cracking surface. These phenomena agree with the findings of Ye et al. (2012) and Kato et al. (2005). In addition, Ye et al. (2013) found that chloride ion concentration increasing with increasing the distance from exposure surface and cracking surface is inaccurate. Ye et al.'s (2013) experiment revealed the importance of mix design during the calculation of chloride diffusion coefficient of cracking concrete, although the influence of mix design is incomparable to the influence of cracking width. The results of Liu et al.'s (2015) model indicated that cracking width played an essential role in controlling the chloride ion concentration for artificial cracking with different widths at a same depth. For flexural cracking, both of cracking width and cracking depth determined the diffusion depth of chloride ions.

7.5.3.3 Effect of Cracking Tortuosity, Orientation, and Density

Test results of Wang et al. (2016a) show that tortuosity of cracking has unremarkable influence on the permeability of chloride ions, when cracking width is less than 150 μm or more than 370 μm. In other words, tortuosity of cracking is not the determinant factor that controls chloride diffusion in wide cracking (>370 μm) and narrow cracking (<150 μm). On the other hand, in cracking with width in the range of 150–370 μm, permeability increases with the increase of tortuosity. Chloride ions could more rapidly penetrate into concrete in planes parallel to the loading direction, because cracking tend to form in the same direction.

Cracking density express the surface area of cracking in every unit-observed area. With a same cracking density, the chloride diffusion coefficient of the primary cracking is far greater than others, indicating the remarkable contribution of the primary cracking to the bulk chloride diffusion coefficient (Wang et al. 2016a). The cracking density in planes parallel to the loading direction is always lower than the cracking density in planes perpendicular to the loading direction (Wang et al. 2016a; Zhou et al. 2012). That is to say, the plane perpendicular to the loading direction crosses the most cracking, reflecting that the development of most cracking are parallel to the loading direction. In fact, the diffusion depth of chloride ions is mainly determined by the effective length of cracking parallel to the diffusion direction (Wang et al. 2016a). Thus, it is necessary to amend the cracking density according to the cracking tortuosity or direction, during the evaluation of chloride diffusion.

7.5.4 Cracking Effect under Different Loading Level

Wang et al. (2016b) determined the D_{nssm} of concrete with and without loading. It was found that all concrete under 75% of ultimate loading exhibited similar behavior with relative significant changes in D_{nssm}, while similar D_{nssm} was obtained between under the loading level of 25% of ultimate loading and no loading. Under 50% of ultimate loading, the D_{nssm} is approximately 1.43 times of the D_{nssm} of non-loading concrete, while this number is 2.24 for concrete under 75% ultimate loading. According to these correlations between loaded concrete and non-loading concrete, it is possible to estimate D_{nssm} of concrete under any loading level and adjust the corresponding service life prediction.

Kurumatani et al. (2017) proposed a stress-strain curve based on a concrete damage model and utilized the stress-strain curve to characterize the chloride diffusion coefficient of damaged concrete, as shown in Figure 7.16. Assuming the material is initially linearly plastic without any damage and the corresponding chloride diffusion coefficient is k_0, the chloride diffusion coefficient would increase as the concrete damage proceeds and eventually converge to k_1. In reality, it is likely that the diffusion coefficient approaches that of bulk water with expanding cracking. This means that the diffusion coefficient cannot infinitely rise as the cracking propagates. Therefore, a damage factor D_e that converges to 1 as the cracking displacement increases was introduced to consider the influence of cracking on the chloride diffusion coefficient. Although parameter k_0 can be measured by experiments, parameter k_1 is difficult to confirm in condition of complete rupture. So, the fracture behavior that relates to the confirmation of k_1 is also characterized by the concrete damage model. In this regard, the diffusion coefficient k_1 can be defined by the apparent concrete diffusion coefficient k_0.

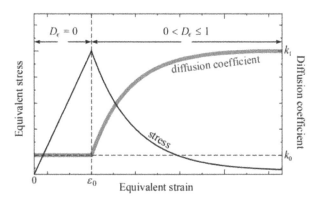

Figure 7.16 Change of diffusion coefficient as damage increases. (From Kurumatani et al. 2017.)

7.6 SUMMARY

This chapter elaborates on the influence of ion type, ion concentration, pore structure, ITZ, chloride adsorption, and cracking on the transport of chloride ions in saturated cement-based materials. The conclusions can be drawn as follows:

(1) The Nernst–Planck equation from single-ion model is not suitable to describe the practical chloride ion transport in cement-based materials, while the Poisson–Nernst–Planck equation, considering interactions between ions, is more suitable to describe the chloride ion transport. But the solving of Poisson–Nernst–Planck equation is too complicated, which limits its application.

(2) Interaction between ions includes the chemical potential between ions, lagging movement of cations, and interaction between double electrical layer and ion clouds. The interaction between ions leads to the dependence of the chloride diffusion coefficient to concentration, as well as time and location. Different types of cation also result in different chloride diffusion coefficients.

(3) Chloride diffusion coefficient increases as porosity and pore diameter increase. ITZ possesses greater chloride diffusion coefficients than cement-based matrix. Although the addition of aggregate introduces ITZ, the bulk chloride diffusion coefficient of cement-based materials is not increased by ITZ because of the dilution and distortion effect of aggregates. In the vast majority of cases, the addition of aggregates decreases chloride diffusion coefficient. However, ITZ has a chance to increase chloride diffusion coefficient when the porosity of the cement-based matrix is very low. Therefore, it is necessary to consider the coupling effect of pore structure and ITZ, when both of them vary.

(4) A different distribution of chloride ion concentration can be obtained when chloride binding and adsorption is considered. It was reported that Friedel's salt was unstable when the pH was lower than 12. In fact, even in the condition of pH > 12 and stable Friedel's salt, it is not reasonable to only consider free chloride transport, because of the transportable physically bound chloride in the diffuse layer. It is worth noting that it may be more accurate to use transportable chloride concentration than free chloride concentration to evaluate durability of concrete exposed to chloride environment.

(5) In addition to cracking width and depth, cracking direction, cracking tortuosity and cracking shrinkability, cracking density is also an essential parameter that influences the chloride diffusion coefficient in concrete. Even the roughness of cracking surface could impact the chloride diffusion coefficient. However, it is still a challenge warranting more studies to establish a model to take these parameters into consideration and correlate them with the determination of chloride diffusion coefficient in concrete.

BIBLIOGRAPHY

Ababneh A, Benboudjema F, Xi Y. Chloride penetration in nonsaturated concrete. *Journal of Materials in Civil Engineering.* 2003, 15:183–191.

Achari G, Chatterji S, Joshi RC. Evidence of the concentration dependent ionic diffusivity through saturated porous media. In: Nilsson L-O, Ollivier JP (eds.), *Chloride Penetration Into Concrete,* Rilem Publications, France; 1995, 74–76.

Aitcin PC, Mehta PK. Effect of coarse aggregate characteristics on mechanical properties of high-strength concrete. *ACI Materials Journal.* 1990, 87:103–107.

Akhavan A, Shafaatian SMH, Rajabipour F. Quantifying the effects of crack width, tortuosity, and roughness on water permeability of cracked mortars. *Cement and Concrete Research.* 2012, 42:313–320.

Aldea CM, Shah SP, Karr A. Effect of cracking on water and chloride permeability of concrete. Journal of Materials Civil Engineer. 1999, 11:181–197.

Andrade C. Calculation of chloride diffusion coefficients in concrete from ionic migration measurement. *Cement and Concrete Research.* 1993, 23:724–742.

Arsenault J, Bigas JP, Ollivier JP. Determination of chloride diffusion coefficient using two different steady state methods: Influence of concentration gradient. Proceedings of the RILEM International Workshop on Chloride Penetration into Concrete, St. Rémy-lès-Chevreuse, French, 1995, 150–160.

Atkins PW. *Physical Chemistry,* 5th ed. Oxford University Press, Oxford; 1994.

Barbarulo R, Marchand J, Snyder KA, Prene S. Dimensional analysis of ionic transport problems in hydrated cement systems: Part 1. Theoretical considerations. *Cement and Concrete Research.* 2000, 30:1955–1960.

Bard AJ, Faulkner LR. *Electrochemical Methods, Fundamental and Applications,* 2nd ed. John Wiley and Sons, Inc.; 2001.

Basheer L, Kropp J. Assessment of the durability of concrete from its permeation properties: A review. *Construction and Building Materials.* 2001, 15:93–103.

Bentz DP, Stutzman PE, Garboczi EJ. Experimental and simulation studies of the interfacial zone in concrete. *Cement and Concrete Research.* 1992, 22:891–902.

Bigas JP. Diffusion of chloride ions through mortars. Doctoral thesis, INSA-Genie Civil, LMDC, INSA de Toulous, France, 1994 (in French).

Birnin-Yauri UA, Glasser FP. Friedel's salt, Ca2Al(OH)6(Cl,OH).2H2O: Its solid solutions and their role in chloride binding. *Cement and Concrete Research.* 1998, 28:1713–1723.

Bockris JOM, Reddy AKN. *Modem Electrochemistry,* 3rd ed. Plenum Press, New York; 1977.

Bockris JOM, Reddy AKN. *Modern Electrochemistry: An Introduction to an Interdisciplinary Area.* Plenum Press, New York; 1970.

Chatterji S. Transportation of ions through cement based materials-Part 1: Fundamental equations and basic measurement techniques. *Cement and Concrete Research.* 1994, 24(5):907–912.

Choi YC, Park B, Pang G-S, Lee K-M, Choi S. Modelling of chloride diffusivity in concrete considering effect of aggregates. *Construction and Building Materials.* 2017, 136:81–87.

Crank J. *The Mathematics of Diffusion,* 2nd ed. Oxford University Press, London; 1975.

Dai JG, Akira Y, Wittmann FH, Yokota H, Zhang P. Water repellent surface impregnation for extension of service life of reinforced concrete structures in marine environments: The role of cracks. *Cement and Concrete Composites*. 2010, 32:101–109.

Debye D, Htickel E, Zeitschrift fur Physik. 1921, 24.

Delagrave A, Bigas JP, OIlivier JP, Marchand J, Pigeon M. Influence of the interfacial zone on the chloride diffusivity of mortars. *Advanced Cement Based Materials*. 1997; 5:86–92.

Djerbi A, Bonnet S, Khelidj A, Baroghel-bouny V. Influence of traversing crack on chloride diffusion into concrete. *Cement and Concrete Research*. 2008, 38:877–883.

Duchesne J, Reardon EJ. Measurement and prediction of portlandite solubility in alkali solutions. *Cement and Concrete Research*. 1995, 25:1043–1053.

Elakneswaran Y, Iwasa A, Nawa T, Sato T, Kurumisawa K. Ion-cement hydrate interactions govern multi-ionic transport model for cementitious materials. *Cement and Concrete Research*. 2010, 40:1756–1765.

Farran J. Contribution minéralogique à l'étude de l'adhérence entre les constituants hydratés des ciments et les matériaux enrobés: Centre d'Etudes et de Recherches de l'Industrie des Liants Hydrauliques; 1956.

Francois R, Toumi A, Ismail M, Castel A, Vidal T. Effect of cracks on local diffusion of chloride and long-term corrosion behavior of reinforced concrete members. Proceedings of the International Workshop on Durability of Reinforced Concrete Under Combined Mechanical and Climatic Loads, Qingdao, China, 2005, 113–122.

Frey R, Balongh T, Balazs GL. Kinetic method to analyse chloride diffusion in various concrete. *Cement and Concrete Research*. 1994, 24:863–873.

Friedmann H, Amiri O, Aït-Mokhtar A. Physical modeling of the electrical double layer effects on multispecies ions transport in cement-based materials. *Cement and Concrete Research*. 2008, 38:1394–1400.

Friedmann H, Amiri O, Ait-Mokhtar A. Shortcomings of geometrical approach in multi-species modelling of chloride migration in cement-based materials. *Magazine of Concrete Research*. 2008, 60:119–124.

Frizon F, Lorente S, Ollivier JP, Thouvenot P. Transport model for the nuclear decontamination of cementitious materials. *Computational Materials Science*. 2003, 27:507–516.

Galus Z. *Fundamentals of Electrochemical Analysis*, 2nd ed. Ellis Horwood, New York; 1994.

Garboczi EJ. Permeability, diffusivity, and microstructural parameters: A critical review. *Cement and Concrete Research*. 1990, 20(4):591–601.

Garboczi EJ, Bentz DP. Computer simulation of the diffusivity of cement-based materials. *Journal of Materials Science*. 1992, 27(8):2083–2092.

Genin JM. On the corrosion of reinforcing steels in concrete in the presence of chlorides. *Materiales de Construction*. 1986, 36:5–16.

Gérard B, Reinhardt HW, Breysse D. Measured transport in cracked concrete. In: H.W. Reinhardt (ed.), *Penetration and Permeability of Concrete: Barriers to Organic and Contaminating Liquids*, E&FN SPON, UK; 1997, 265–324.

Gjarv OE, Vennesland O. Evaluation and control of steel corrosion in offshore concrete structures, concrete durability. In: Scanlon JM (ed.), Proceedings of the Katharine and Bryant Mather International Symposium, ACI SP-100; 1987, Vol. 2, 1575–1602.

Gjorv OE, Vennesland O. Diffusion of chloride ions from seawater into concrete. *Cement and Concrete Research.* 1979, 9:229–238.

Goto S, Tsuuetani M, Yanagida H, Kondo R. Diffusion of chloride ion in hardened cement paste. *Yogyo Kyokaishi.* 1979, 87:126–133.

Gowripalan N, Sirivivatnanon V, Lim CC. Chloride diffusivity of concrete cracked in flexure. *Cement and Concrete Research.* 2000, 30:725–730.

He F, Shi C, Hu X, Wang R, Shi Z, Li Q, Li P, An X. Calculation of chloride ion concentration in expressed pore solution of cement-based materials exposed to a chloride salt solution. *Cement and Concrete Research.* 2016, 89:168–176.

Hidalgo A, Andrade C, Goñi S, Alonso C. Single ion activities of unassociated chlorides in NaCl solutions by ion selective electrode potentiometry. Proceedings of the 10th International Congress on the Chemistry of Cement, Gothenburg; 1997, Vol. IV, 9.

Ishida T, Iqbal P, Anh H. Modeling of chloride diffusivity coupled with non-linear binding capacity in sound and cracked concrete. *Cement and Concrete Research.* 2009, 39:913–923.

Ismail M, Toumi A, François R, GagnéR. Effect of crack opening on the local diffusion of chloride in cracked mortar samples. *Cement and Concrete Research.* 2008, 38:1106–1111.

Jang SY, Kim BS, Oh BH. Effect of crack width on chloride diffusion coefficients of concrete by steady-state migration tests. *Cement and Concrete Research.* 2011, 41:9–19.

Jiang L, Song Z, Yang H, Pu Q, Zhu Q. Modeling the chloride concentration profile in migration test based on general Poisson–Nernst–Planck equations and pore structure hypothesis. *Construction and Building Materials.* 2013, 40:596–603.

Johannesson B, Yamada K, Nilsson LO, Hosokawa Y. Multi-species ionic diffusion in concrete with account to interaction between ions in the pore solution and the cement hydrates. *Materials and Structures.* 2007, 40:651–665.

Justnes H, Rodum E. Chloride ion diffusion coefficients for concrete—A review of experimental methods. Proceedings of the 10th International Congress on the Chemistry of Cement, Gothenburg, 1997, Vol. IV, 8.

Kato E, Kato Y, Uomoto T. Development of simulation model of chloride ion transportation in cracked concrete. *Journal of Advanced Concrete Technology.* 2005, 3:85–94.

Khitab A, Lorente S, Olivier JP. Predictive model for chloride penetration through concrete. *Magazine of Concrete Research.* 2005, 57:511–520.

Krabbenhoft K, Krabbenhoft J. Application of the Poisson–Nernst–Planck equations to the migration test. *Cement and Concrete Research.* 2008, 38:77–88.

Kubo J, Sawada S, Page CL. Page MM. Electrochemical injection of organic corrosion inhibitors into carbonated cementitious materials: Part 2. Mathematical modelling. *Corrosion Science.* 2007, 49:1205–1227.

Kurumatani M, Anzo H, Kobayashi K, Okazaki S, Hirose S. Damage model for simulating chloride concentration in reinforced concrete with internal cracks. *Cement and Concrete Composites.* 2017, 84:62–73.

Li C, Jiang L, Xu N, Jiang S. Pore structure and permeability of concrete with high volume of limestone powder addition. *Powder Technology.* 2018, 338:416–424.

Li L, Page CL. Finite element modelling of chloride removal from concrete by an electrochemical method. *Corrosion Science.* 2000, 42:2145–2165.

Li L, Page CL. Modelling and simulation of chloride extraction from concrete by using electrochemical method. *Computational Materials Science*. 1998, 9:303–308.

Little DR (ed.). *CRC Handbook of Chemistry and Physics*, 76th ed. CRC Press, Boca Raton, FL; 1996.

Liu QF. Multi-phase modelling of multi-species ionic migration in concrete. PhD thesis, University of Plymouth; 2014.

Liu Q, Li LY, Easterbrook D, Yang J. Multi-phase modelling of ionic transport in concrete when subjected to an externally applied electric field. *Engineering Structures*. 2012, 42:201–213.

Liu Q, Easterbrook D, Yang J, Li L. A three-phase, multi-component ionic transport model for simulation of chloride penetration in concrete. *Engineering Structures*. 2015, 86:122–133.

Liu Q, Yang J, Xia J, Easterbrook D, Li L, Lu X. A numerical study on chloride migration in cracked concrete using multi-component ionic transport models. *Computational Materials Science*. 2015, 99:396–416.

Lorente S, Carcasses M, Ollivier JP. Penetration of ionic species into saturated porous media: The case of concrete. *International Journal of Energy Research*. 2003, 27:907–917.

MacDonald KA, Northwood DO. Experimental measurements of chloride ion diffusion rates using a two-compartment diffusion cell: Effects of material and test variables. *Cement and Concrete Research*. 1995, 25:1407–1416.

Maekawa K, Ishida T, Kishi, T. Multi-scale modeling of concrete performance integrated material and structural mechanics. *Journal of Advanced Concrete Technology*. 2003, 1:91–126.

Marchand J. Modeling the behavior of unsaturated cement systems exposed to aggressive chemical environments. *Materials and Structures*. 2001, 34:195–200.

Marchand J, Gérard B, Delagrave A. Ions transport mechanisms in cement-based materials, Report GCS-95-07, Department of Civil Engineering, University of Laval, Québec, Canada; 1995.

Marsavina L, Audenaert K, Schutter G, Faur N, Marsavina D. Experimental and numerical determination of the chloride penetration in cracked concrete. *Construction and Building Materials*. 2009, 23:264–274.

Martin-Perez B, Zibara H, Hooton RD. A Study of the effect of chloride binding on service life predictions. *Cement and Concrete Research*. 2000, 30:1215–1223.

Mohammed MK, Dawson AR, Thom NH. Macro/micro-pore structure characteristics and the chloride penetration of self-compacting concrete incorporating different types of filler and mineral admixture. *Construction and Building Materials*. 2014, 72:83–93.

Mu S, Schutter GD, Ma BG. Non-steady state chloride diffusion in concrete with different crack densities. *Materials and Structures*. 2013, 46(1):123–133.

Narsillo GA, Li R, Pivonka P, Smith DW. Comparative study of methods used to estimate ionic diffusion coefficients using migration tests. *Cement and Concrete Research*. 2007, 37:1152–1263.

NTBuild 443. *Concrete, Hardened: Accelerated Chloride Penetration*. Nordtest, Finland; 1995, 1–5.

Olivier T. Prediction of chloride penetration into saturated concrete—Multi-species approach: Ph.D. thesis, Department of Building Materials, Chalmers University of Technology, Goteborg, Sweden; 2000.

Pankow JF. *Aquatic Chemistry Concepts*. Lewis; 1994.

Philip H. *Electrochemistry*, 2nd ed. Chapman & Hall, New York; 1994.

Pivonka P, Hellmich C, Smith D. Microscopic effects on chloride diffusivity of cement pastes—A scale-transition analysis. *Cement and Concrete Research*. 2004, 34:2251–2260.

Reardon EJ. An ion interaction model for determination of chemical equilibria in cement/water systems. *Cement and Concrete Research*. 1990, 20:175–192.

Rodriguez OG, Hooton RD. Influence of cracks on chloride ingress into concrete. *ACI Materials Journal*. 2003, 100:120–126.

Roy DM, Kumar A, Rhodes JP. Diffusion of chloride and cesium ions in Portland cement pastes and mortars containing blast furnace slag and fly ash. Proceedings of 2nd International Conference on the Use of Fly Ash, Silica Fume, Slag and Natural Pozzolans in Concrete, Madrid, ACI SP-91, 1986, 1423–1444.

Saetta AV, Scotta RV, Vitaliani RV. Analysis of chloride diffusion into partially saturated concrete. *Materials Journal*. 1993, 90:441–451.

Sahmaran M. Effect of flexure induced transverse crack and self-healing on chloride diffusivity of reinforced mortar. *Journal of Materials Science*. 2007, 42:9131–9136.

Samson E, Marchand J, Beaudoin J. Modeling the influence of chemical reactions on the mechanisms of ionic transport in porous materials: An overview. *Cement and Concrete Research*. 2000, 30:1895–1902.

Samson E, Marchand J, Snyder KA. Calculation of ionic diffusion coefficients on the basis of migration test results. *Materials and Structures*. 2003, 36:156–165.

Scrivener KL, Crumbie AK, Laugesen P. The interfacial transition zone (ITZ) between cement paste and aggregate in concrete. *Interface Science*. 2004, 12:411–421.

Snyder KA, Marchand J. Effect of speciation on the apparent diffusion coefficient in nonreactive porous systems. *Cement and Concrete Research*. 2001, 31:1837–1845.

Spiesz P, Ballari MM, Brouwers HJH. RCM: A new model accounting for the non-linear chloride binding isotherm and the non-equilibrium conditions between the free- and bound-chloride concentrations. *Construction and Building Materials*. 2012, 27:293–304.

Sun G, Zhang Y, Sun W, Liu Z, Wang C. Multi-scale prediction of the effective chloride diffusion coefficient of concrete. *Construction and Building Materials*. 2011a, 25:3820–3831.

Sun G, Sun W, Zhang Y, Liu Z. Relationship between chloride diffusivity and pore structure of hardened cement paste. *Journal of Zhejiang University-Science A (Applied Physics and Engineering)*. 2011b, 12:360–367.

Suryavanshi AK, Swamy RN. Stability of Friedel's salt in carbonated concrete structural elements. *Cement and Concrete Research*. 1996, 26:729–741.

Tang L, Nilsson L. Rapid determination of the chloride diffusivity in concrete by applying an electrical field. *ACI Materials Journal*. 1992, 89:49–53.

Tang L. Chloride transport in concrete—Measurement and prediction. Doctoral thesis, Department of Building Materials, Chalmers Universities of Technology, Publication P-96:6, Gothenburg, Sweden; 1996.

Tang L. Concentration dependence of diffusion and migration of chloride ions— Part 1: Theoretical considerations. *Cement and Concrete Research.* 1999a, 29:1463–1468.

Tang L. Concentration dependence of diffusion and migration of chloride ions— Part 2. Experimental evaluations. *Cement and Concrete Research.* 1999b, 29:1469–1474.

Tian Y, Tian Z, Jin N, Jin X, Yu W. A multiphase numerical simulation of chloride ions diffusion in concrete using electron microprobe analysis for characterizing properties of ITZ. *Construction and Building Materials.* 2018, 178:432–444.

Toumi A, Francois R, Alvarado O. Experimental and numerical study of electrochemical chloride removal from brick and concrete specimens. *Cement and Concrete Research.* 2007, 37:54–62.

Truc O, Ollivier JP, Nilsson LO. Numerical simulation of multi-species transport through saturated concrete during a migration test-msdiff code. *Cement and Concrete Research.* 2000a, 30:1581–1592.

Truc O, Ollivier JP, Nilsson LO. Numerical simulation of multi-species diffusion. *Materials and Structures.* 2000b, 33:566–573.

Ushiyama H, Goto S. Diffusion of various ions in hardened portland cement pastes. 6th International Congress on the Chemistry of Cement, Moscow, 1974, Vol. II, 331–337.

Ushiyama H, Iwakakura H, Fukunaga T. Diffusion of Sulphate in Hardened Portland Cement, Cement Association of Japan, Review of 30th General Meeting, 1976, 47–49.

Wang H, Dai J, Sun X, Zhang X. Characteristics of concrete cracks and their influence on chloride penetration. *Construction and Building Materials.* 2016a, 107:216–225.

Wang J, Muhammed Basheer PA, Nanukuttan Sreejith V, Long Adrian E, Bai Y. Influence of service loading and the resulting micro-cracks on chloride resistance of concrete. *Construction and Building Materials.* 2016b, 108:56–66.

Wang K, Jansen D, Shah SP, Karr A. Permeability study of cracked concrete. *Cement and Concrete Research.* 1997, 27:381–393.

Wang Y, Li LY, Page CL. A two-dimensional model of electrochemical chloride removal from concrete. *Computational Materials Science.* 2001, 20:196–212.

Wang Y, Wu L, Wang Y, Liu C, Li Q. Effects of coarse aggregates on chloride diffusion coefficients of concrete and interfacial transition zone under experimental drying-wetting cycles. *Construction and Building Materials.* 2018, 185:230–245.

Witting D. Rapid measurement of chloride permeability of concrete. *Public Roads.* 1981, 45:101–112.

Wittmann FH, Zhao T, Ren Z, Guo P. Influence of surface impregnation with silane on penetration of chloride into cracked concrete and on corrosion of steel reinforcement. *International Journal of Modelling, Identification and Control.* 2009, 7:135–141.

Xi Y, Bažant ZP. Modeling chloride penetration in saturated concrete. *Journal of Materials in Civil Engineering.* 1999, 11:58–65.

Xia J, Li L. Numerical simulation of ionic transport in cement paste under the action of externally applied electric field. *Construction and Building Materials*. 2013, 39:51–59.

Yang C, Su J. Approximate migration coefficient of interfacial transition zone and the effect of aggregate content on the migration coefficient of mortar. *Cement and Concrete Research*. 2002, 32:1559–1565.

Yang C. On the relationship between pore structure and chloride diffusivity from accelerated chloride migration test in cement-based materials. *Cement and Concrete Research*. 2006, 36:1304–1311.

Yang C, Cho S, Wang L. The relationship between pore structure and chloride diffusivity from ponding test in cement-based materials. *Materials Chemistry and Physics*. 2006, 100:203–210.

Ye H, Jin N, Jin X, Fu C. Model of chloride penetration into cracked concrete subject to drying-wetting cycles. *Construction and Building Materials*. 2012, 36:259–269.

Ye H, Tian Y, Jin N, Jin X, Fu C. Influence of cracking on chloride diffusivity and moisture influential depth in concrete subjected to simulated environmental conditions. *Construction and Building Materials*. 2013, 47:66–79.

Yu S, Page CL. Computer simulation of ionic migration during electrochemical chloride extraction from hardened concrete. *British Corrosion Journal*. 1996, 31:73–75.

Zhang J, Buenfeld NR. Presence and possible implications of a membrane potential in concrete exposed to chloride solution. *Cement and Concrete Research*. 1997, 27:853–859.

Zhang T. Chloride diffusivity in concrete and its measurement from steady state migration testing, Doctoral thesis. TrondheimUniversity of Science and Technology, Norway; 1997, 132.

Zhang T, Gjørv OE. Diffusion behavior of chloride ions in concrete. *Cement and Concrete Research*. 1996, 26:907–917.

Zhang T, Gjarv OE. Effect of ionic interaction in migration testing of chloride diffusivity in concrete. *Cement and Concrete Research*. 1995, 25:1535–1542.

Zheng J-J, Zhang J, Zhou X-Z, Song W-B. A numerical algorithm for evaluating the chloride diffusion coefficient of concrete with crushed aggregates. *Construction and Building Materials*. 2018, 171:977–983.

Zheng J-j, Wong HS, Buenfeld NR. Assessing the influence of ITZ on the steady-state chloride diffusivity of concrete using a numerical model. *Cement and Concrete Research*. 2009, 39:805–813.

Zheng J, Zhou X. Effective medium method for predicting the chloride diffusivity in concrete with ITZ percolation effect. *Construction and Building Materials*. 2013, 47:1093–1098.

Zheng J, Zhou X, Wu Y, Jin X. A numerical method for the chloride diffusivity in concrete with aggregate shape effect. *Construction and Building Materials*. 2012, 31:151–156.

Zhou C, Li K, Pang X. Effect of crack density and connectivity on the permeability of microcracked solids. *Mechanics of Materials*. 2011, 43:969–978.

Zhou C, Li K, Pang X. Geometry of crack network and its impact on transport properties of concrete. *Cement and Concrete Research*. 2012, 42(9):1261–1272.

Chapter 8

Simulation and Modeling of Chloride Transport in Cement-Based Materials

8.1 INTRODUCTION

Concrete structures are a kind of goods which should be in good performance during its service life. The performance of concrete structures usually decays with time, due to various environmental actions. During its service period, maintenance may be applied to the structures to restore the performance of the structures. When the performance of the structures decays to beyond the acceptance limit, the structures have to be demolished or discarded. The period from the beginning to end use is defined as the service life. It is critical to know the service life of structures, so the investor or decision maker can do the life-cycle cost analysis and make the decision to invest in the structures or not. This is especially important for key infrastructures. For instance, the service life of Hong Kong–Zhuhai–Macao bridge is 120 years, and the investment of the bridge is more than 12 billion renminbi (RMB). Service life prediction models have been developed to predict the service life of bridges in marine environments (Li 2016).

In addition, much of the world's concrete infrastructure was built more than 50 years ago. Many of the aged structures are showing signs of degradation. The growing concerns of degrading concrete structures has led to an increased interest in the durability of cementitious materials around the world. In-depth analyses of the various deterioration mechanisms at the microstructure/chemical level have contributed to our global understanding of the underlying degradation phenomena. In the meantime, great efforts have been made to model the various degradation of concrete structures at particular environmental exposure. Precise modeling can be used to assist engineers in assessing the repair/maintenance solutions to extend the service life of existing structures.

Therefore, to predict the service life based on the selected materials and structure design is necessary for life cycles' cost of new infrastructure or maintenance strategy for old infrastructures. However, since the deterioration of concrete in different environments is quite different and very complicated, it is very difficult to predict the service life of concrete structures precisely. For concrete structures in marine environments, the corrosion of reinforced steel is the major concern. It may intertwine with the carbonation, abrasion, sulfate attack, and freezing–thawing cycles. Thus, it greatly increases the difficulty of modeling. However, to predict the service life of concrete in marine environments based on chloride transport in concrete is generally accepted by academia and industry.

In the context of chloride-induced corrosion of reinforced steel, it is reasonable to consider the service life of reinforced concrete structures as consisting of two stages (Tuutti 1982b), as shown in Figure 8.1. The first stage is generally related to the time needed for a critical chloride concentration to reach the steel bar, which is related to the transport of chloride ion in concrete. In some service life prediction models (Magge et al. 1996; Boddy et al. 1999), the first stage, i.e., the initiation period, is regarded as the service life. The subsequent propagation stage extends on to the time when the structure has been damaged by corrosion beyond acceptable limits, which is related to the expansion of corrosion products in concrete and the resulting cracks. Some models also include this period in the service life (Maaddawy and Soudki 2007). Once the first stage is completed, the second stage will propagate relatively fast. Over the past decades, many models, from oversimplified to overcomplicated, have been developed around the world.

It is worth mentioning that Angst (2019) calculated the corrosion initiation time of steel-reinforced concrete, utilizing input data derived from both laboratory specimens and from structures, illustrating the poor predictive power of state-of-the-art models. It is generally assumed that improving chloride ingress models will improve the prediction of the time-to-corrosion. However, Angst (2019) showed that the bottle neck to more reliable predictions are rather (i) the lack of fundamental understanding of corrosion initiation, (ii) the use of non-representative laboratory results, and (iii)

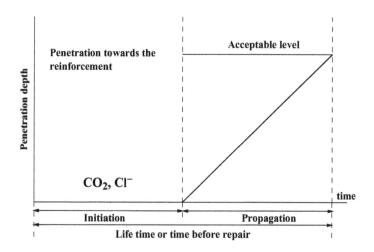

Figure 8.1 Schematic diagram of service life model of reinforced concrete.

ignoring the size-effect in localized corrosion. This is far more complicated, and beyond the scope of this chapter. This chapter still uses the model by Tuutti (1982b), and only reviews the models related to the chloride transport in concrete.

As stated above, the transport of chloride in concrete is closely related to the water saturation degree of concrete. Thus, the modeling of chloride in saturated and unsaturated cement-based materials are discussed separately. In addition, the major standards or codes related with chloride-induced durability of concrete structures are also introduced in this chapter.

8.2 MODELING CHLORIDE TRANSPORT IN SATURATED CONCRETE

Based on the nature of the model parameters, the model can be divided into deterministic and probabilistic models. In a deterministic model, the model parameters are normally fixed, whereas in a probabilistic model the model parameters are treated as continuous stochastic variables, which are characterized by a mean value, a standard deviation, and a probabilistic density function. Figure 8.2 (Gulikers 2007) illustrates the difference between probabilistic methods and deterministic methods. According to deterministic methods, corrosion occurs if C (the chloride content at the steel surface) exceeds C_{crit}, and no corrosion occurs if C is lower than C_{crit}. on The other hand, in the probabilistic model, when C is lower than $C_{crit,min}$, there is no corrosion; when $C_{crit,min} < C < C_{crit,max}$, there is an increasing probability of corrosion initiation. When $C > C_{crit,max}$, corrosion occurs certainly. Cao et al. (2019) reviewed the extensive body of literature on C_{crit} published in Chinese since the 1960s. It was found that C_{crit} scatters widely and cannot be predicted on the basis of parameters such as water/binder (w/b) ratio, binder type, steel surface condition, etc. It appears that the stress state of the reinforcing steel may play a more important role than generally assumed. They concluded that the state-of-the-art on C_{crit} has advanced negligibly over the last decades.

In principle, there are two types of model within the deterministic models, i.e., "physical" and "empirical" models. In "physical models" all physical and electrochemical processes are described as scientifically correct as possible. "Physical models" need independently determined data, but not curve-fitting data. The field exposure data is used to validate predictions. If the agreement between predicted results and exposure data is not good enough, the model must be improved, or better data must be determined. In "empirical models," the experimental data in certain periods are fitted to the mathematical model, normally an error function solution to Fick's second law, which results in some empirical parameters with no physical meaning. They are often used to extrapolate the service life of concrete structures. Note that the experimental period is normally much shorter

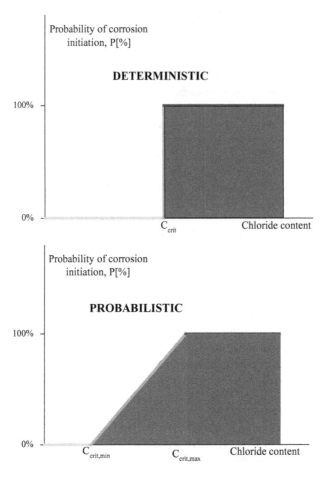

Figure 8.2 Comparison between a probabilistic and a determininistic approach by considering Ccrit as a static and stochastic variable, respectively.

than the service life. Using the early data to extrapolate the future behavior is quite questionable, as shown in Figure 8.3.

Thus, an "empirical model" is only suitable for the structure with the same time period and conditions as the one tested at the same period and conditions. These empirical models based on Fick's second law are now very popular in the scientific and engineering community. However, in spite of the potential higher accuracy of the physical models more complicated than the models based on Fick's law, engineers show reluctance to use this kind of model, and are even less prone to introduce them into national codes or standards. It is worth mentioning that models based on Fick's law or some physical models are theoretically developed on water-saturated concrete. Chloride transport in unsaturated concrete is a much

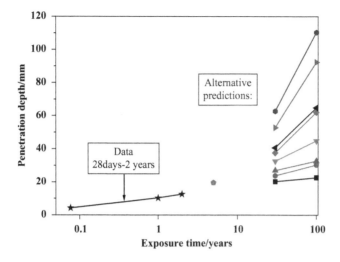

Figure 8.3 The alternative predictions for 30–100 years in figure compared to the observation data between 28 days and 2 years. The circle at 5 years represents later measurements that were not available for the predictions.

more complicated topic. In this section, the general principle of three types of model were first introduced. A detailed description was given to some representative models.

8.2.1 Empirical Model (Fickian Model)

The error function solution to Fick's second law is widely used as the theoretical basis of the service life prediction model. However, it has been noticed for quite a long time that diffusion coefficients change over time; even the surface chloride concentration changes (normally increases) with time. With respect to the surface chloride concentration, Amey et al. (1998) suggested to use linear or power function to describe the change of surface chloride concentration over time. Weyers (1998) investigated the surface chlorides of 15 bridge decks in the snowbelt region at a shallow depth below the surface over a period of 15 years. He found that the chloride concentration increased with time, as shown in Figure 8.4. Weyers argued that surface chloride concentration reaches constant in five years. Compared to the service life of 40 years of the bridge, it is reasonable to use a near constant surface concentration in the solution to the Fick's second law.

Kassir and Ghosn (2002) analyzed Weyers' results and proposed an exponential function for the surface chloride. Kassir obtained a solution to Fick's second law with a varying surface chloride, and found that constant surface chloride may result in 100% conservative estimation compared to

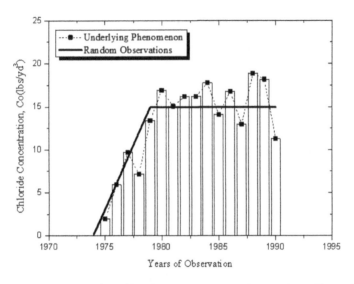

Figure 8.4 The change in surface chloride concentration with time. (From Weyer 1998.)

the varying surface chloride concentration approach. In Weyers' investigation (1998), the concrete structures were subjected to drying–wetting cycles. The drying–wetting cycles could explain the build-up of chloride in the surface. In the submerged zone, the increase of chloride over time was also observed (Tang and Lars-Olof 2000). Tang (2008) ascribed it to the increase of bound chloride over time, while the free chloride in the pore solution remains relatively constant.

Pang et al. (2016) proposed that the C_s increased non-linearly with the exposure time, remaining constant after 20 years.

Shakouri and Trejo (2017) proposed an improved time-variant C_s model which is hypothesized using general physical concepts and is then validated by an empirical study. The proposed model, as opposed to the existing time-variant models, not only accounts for the variability of C_s with exposure time but also incorporates the effects of time to exposure and the effects of the concentration of chlorides in the exposure environment. The model assumes that C_s is sigmoidal in shape with an asymptote that is a function of the concentration of chlorides in the environment. The input variables for the proposed model were selected based on best subset sampling analysis on the results of the experimental work to determine the influence of water-to-cement ratio, time to exposure, the concentration of chlorides in the exposure environment, and exposure time on C_s. The accuracy of the proposed model is assessed versus existing time-variant C_s models, and the results indicate that the proposed model better predicts the C_s in concrete exposed to chlorides.

Shakouri and Trejo (2018) found w/c does not have a significant influence on the maximum C_s; an increase in w/c results in a significant increase in the depth at which the maximum chloride concentration occurs (Δx). Shakouri and Trejo (2018) believed that it potentially impacts on the numerical estimation of the apparent chloride diffusion coefficient (D_a) and surface chloride concentration (C_s). Moreover, while exposure duration has a significant influence on C_{max}, it does not influence Δx. The concentration of chlorides in the exposure environment and the time-to-exposure have opposite effects on Δx and the maximum C_s. While an increase in time-to-exposure results in an increase in Δx, it results in a reduction in the maximum C_s at later ages. Similarly, while Δx reduces with an increase in the concentration of chlorides, the maximum C_s increases nonlinearly with an increase in the concentration of chlorides. The concrete skin can sometimes have a thickness that goes beyond the first millimeter from the concrete surface. When a chloride profile shows the maximum phenomenon, care must be taken in using the estimated D_a and C_s values in service life prediction models.

Owing to the ongoing hydration of cementing materials, permeability of concrete decreases over time. Takewake and Mastumoto (1988) are probably the first who pointed out the dependency of the chloride diffusion coefficient D on the exposure period t and used a purely empirical equation to describe the decrease of diffusion coefficient with time. Compared to the time dependency of surface chloride, more attention was given to the time dependency of diffusion coefficient with time, which is often written as follows (Mangat and Molloy 1994; Maage et al. 1996; Duracrete 1998; Stanish and Thomas 2003; Nokken et al. 2006; Tang 2008):

$$D_i = D_{ref}\left(\frac{t_{ref}}{t_i}\right)^m \qquad (8.1)$$

where D_{ref} and D_i is the diffusion coefficient at t_{ref} and t_i; m is a constant, which depends on the binder of concrete.

There are solutions to Fick's second law where either surface chloride changes over time or diffusion coefficient changes over time (Crank 1975). No solutions to Fick's second law are given in this book (Crank 1975) that combines a time-dependent surface chloride concentration and a time-dependent diffusion coefficient. Yu (2004) solved this equation with different boundaries. The correctness of his solution still needs to be proved. At present, most applied empirical models are only based on the time-dependent diffusion coefficient. There were some solutions to Fick's second law with a time-dependent diffusion coefficient, as the solutions reported in the literature (Mangat and Molloy 1994; Magne et al. 1996; Duracete 1998). However, this solution is regarded as incorrect. The correct solution was presented a few years later by some authors (Stanish and Thomas 2003; Tang 2007):

$$\frac{c}{c_s} = 1 - \mathrm{erf}\left(\frac{x}{2\sqrt{D_{app}t}}\right) = 1 - \mathrm{erf}\left(\frac{x}{2\sqrt{\dfrac{D_{ref}}{1-m}\left[\left(1+\dfrac{t_{ex}}{t}\right)^{1-m} - \left(\dfrac{t_{ex}}{t}\right)^{1-m}\cdot\left(\dfrac{t_{ref}}{t}\right)t\right]}}\right)$$

(8.2)

where t_{ex} is the age of concrete at the start of exposure, t is the duration of exposure, t_{ref} is the reference age, and D_{ref} is the reference diffusion coefficient. Obviously, Equation 8.2 only applies under the implicit assumption that chloride binding is time independent, and free chloride is linearly proportional to bound chloride.

In spite of the oversimplified assumptions, empirical models are widely used by engineers in practice, due to its simple mathematical expressions and theoretical basis. Based on the above discussion, some conclusions can be drawn on the empirical models:

- The use of short-term experimental results to extrapolate long-term performance may be incorrect. New experiments are needed for new structures and new environments.
- The time dependency of the diffusion coefficient attracted a lot of attention, but the time dependency of surface chloride concentration is not well understood. At present, only the time-dependent diffusion coefficient is implemented in the empirical models.

Due to the merits of empirical models, some commercial models have been developed, for example Life-365™. The following will give a detailed description of the model Figure 8.5.

- Life-365™ service life prediction model

In 1998, a workshop, entitled "Models for Predicting Service Life and Life-Cycle Cost of Steel-Reinforced Concrete," was sponsored by the National Institute of Standards and Technology (NIST), ACI, and the American Society for Testing and Materials (ASTM). At this workshop, a decision was made to attempt to develop a "standard model" under the jurisdiction of the existing ACI Committee 365 "Service Life Prediction." Afterwards, a consortium was established under ACI's SDC to fund the development of an initial life-cycle cost model based on the existing service life model developed at the University of Toronto (Boddy et al. 1999; Thomas and

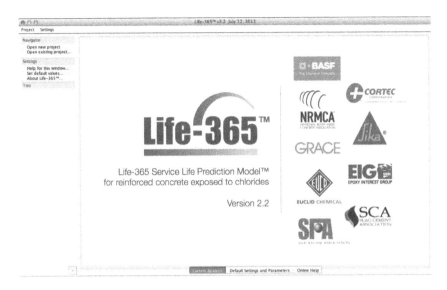

Figure 8.5 Interface of Life-365TM service life prediction software.

Bamforth 1999). The Life-365™ v1.0 program and manual were written by E. C. Bentz and M. D. A. Thomas. The Life-365™ v2.2 program and manual are adaptations of these documents, and were written by M. A. Ehlen under contract to the Life-365 Consortium III, which consists of BASF Admixture Systems, Cortec, Epoxy Interest Group (Concrete Reinforcing Steel Institute), Euclid Chemical, Grace Construction Products, National Ready-Mixed Concrete Association, Sika Corporation, Silica Fume Association, Slag Cement Association. In this model, only the service life because of chloride-induced corrosion is dealt with.

This model is quite versatile. It can be used for prediction of service life of concrete structures subjected to chloride environments, and also can be used for strategic plan of maintenance. It also uses Tuutti's model (Tuutti 1982b). The service life of concrete includes two stages, i.e., the initiation period and propagation period. Before the end of the service life of structures, repair can be made to extend the service life, as shown in Figure 8.6. This model can do life cycle cost analysis and help to make the repair plan. Only the first stage is discussed in this section.

As for the chloride transport in concrete, this model is based on the error function solution to Fick's second law:

$$c(x,t) = c_s \left[1 - \text{erf} \left(\frac{x}{\sqrt{4D_a t}} \right) \right]$$

(8.3)

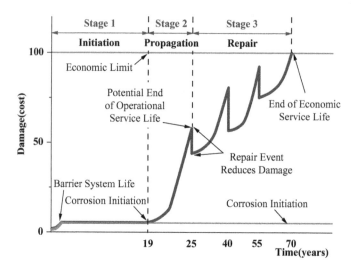

Figure 8.6 The service life of structures with maintenance.

In Equation 8.3, C_s, D_a are input parameters. The value of these two parameters determines the precision of the model.

- Surface chloride concentration (C_s)

As reported by Weyers (1998), surface chloride concentration linearly built up with time, and reached a maximum value after a certain time. The C_s depends on the type of structure, geographic location, and exposure. In this model, the time to reach the maximum C_s is specified as seven years. The C_s can be given by different methods in this model.

- Given by software based on type of structure, geographic location and exposure.
- Directly input by user.
- A calculator based on ASTMC 1556. Chloride profile determined in laboratory according to ASTM C 1556 which is similar to NT Build 443 can be input into the calculator, and the maximum chloride surface concentration can be obtained by curve fitting.

- Diffusion coefficient

Diffusion coefficient increases with time due to cement hydration. Equation 8.1 was adopted in this model. In addition, the temperature dependency of diffusion coefficient was also taken into consideration, as given in Equation 8.4.

$$D(T) = D_{ref} \cdot \exp\left[\frac{U}{R} \cdot \left(\frac{1}{T_{ref}} - \frac{1}{T}\right)\right] \tag{8.4}$$

The temperature profile over one year can be

- Given by software based on meteorological database.
- Directly input by user in terms of monthly average temperature.

Diffusion coefficient D_{28} in this model can be either determined by ASTM C 1556 or calculated by this model based on mix proportions of concrete. The input parameters are w/c ratio and replacement level of supplementary cementing materials. In this model, blank concrete with w/c ratio of 0.4 was taken as reference. Its diffusion coefficient is $D28 = 7.9 \times 10^{-12}$ m²/s at 20°C.

For blank concrete with different w/c ratios

$$D_{28} = 1 \times 10^{(-12.06+2.40 \text{ w/cm})}$$

$m = 0.2$

For silica fume concrete with replacement level <15%

$$D_{SF} = D_{PC} \cdot e^{-0.165 \cdot SF}$$

$m = 0.2$

For fly ash or slag concrete with FA <50% and SG <70%

$$m = 0.2 + 0.4(\%FA/50 + \% SG/70)$$

The advantages and disadvantages were as follows:

- Advantages:
  - It is well developed commercial software.
  - It is very powerful, versatile, and user-friendly.
  - It is semi-empirical-based and based on a large database.
- Disadvantages:
  - The chemical and physical interactions between chloride and concrete are lumped into some empirical parameters.

8.2.2 Physical Models

As stated above, in "physical models" all physical and electrochemical processes are described as scientifically correct as possible, often described by partial differential equations. The analytical solutions to these differential equations are very difficult or impossible to find. Thus, numerical solutions, such as finite element or finite difference methods, are needed to solve the partial differential equations by means of computer. In reality, the transport of chloride in concrete is a very complicated process; it involves moisture transport, heat transfer, etc. Even when diffusion alone is considered, the process is still very complicated. Many researchers (Tang 1996; Masi et al. 1997; Samson et al. 1999d; Truc 2000b; Khitab et al. 2005; Fenaux et al. 2019) have developed different physical models for the diffusion alone case.

In these models, saturated concrete is treated as a two-phase material, i.e., pore solution and solid phase. Chloride only transports in the pore solution. The transport of chloride in solid phase is ignored. Generally speaking, a physical model mainly consists of three parts: the mechanisms described by partial differential equations, inputs, and outputs. The following will give a brief description on some representative physical models.

(1) Model by Tang (ClinConc 1996)
- Equations describing the transport process
  - The mass balance equation
  - The flux equation—Fick's first law
  - Chloride binding relationship—Freundlich isotherm
  - The effect of environmental conditions
  - The effect of material characteristics
- Inputs
  - The non-steady-state migration coefficient which is a function of the depth, time, and temperature, and determined according to NT Build 492
  - The mix proportions (cement, aggregates, water)
  - The environmental conditions (chloride concentration, temperature)
  - The cast conditions of concrete
- Outputs
  - Free chloride profile
  - Total chloride profile

In 1996, Tang proposed this well-thought-out physical model in which many physical and chemical phenomena are taken into account. Finite difference method is used to solve the partial differential equations. However, in the model, chloride ion is treated as a neutral particle, and has no interaction with other species. It is well known that a chloride ion is a negative particle, and the pore solution of concrete is a concentrated solution with ions like K^+, Na^+, Ca^{2+}, OH^-, SO_4^{2-}, etc. When the chloride transports in such a concentrated solution, the interactions between different species should be taken into account. It is worth mentioning that the ClinConc needs numerical iterations by means of computer. This may restrain the engineering applications of the ClinConc. Based on the ClinConc, Tang (2008) proposed a more engineer-friendly expression which doesn't need numerical iterations.

In the ClinConc model, free chloride concentration, instead of total chloride contents, is used in. So, the ingress of chloride into concrete can be described by Equation 8.3. D_a is referred to as the apparent chloride diffusion coefficient that is not a constant but time-dependent:

$$D_a = \frac{D_0}{1-n} \cdot \left(\frac{t'_0}{t}\right)^n \cdot \left[\left(1+\frac{t'_{ex}}{t}\right)^{1-n} - \left(\frac{t'_{ex}}{t}\right)^{1-n}\right] = D_0 \cdot f(n,t) \qquad (8.5)$$

where D_0 is the diffusion coefficient of Fick's second law at age t'_0, n is the age factor, and t'_{ex} is the age of concrete at the start of exposure. Notice that, in this paper, t' denotes the age of concrete and t denotes the duration of exposure.

Tang et al. (2012) found that the measured chloride diffusion coefficient decreased with age. After a certain age, like the development of hydration process, the chloride diffusion coefficient becomes more or less constant according to Tang's laboratory test. Therefore, in the original ClinConc model the time-dependent factor of chloride diffusion coefficient, $f_D(t')$, is described by the following equation:

$$f_D(t') = \begin{cases} \left(\dfrac{t'_{D_s}}{t'}\right)^{\beta_t} & t' < t'_{D_s} \\ 1 & t' \geq t'_{D_s} \end{cases} \qquad (8.6)$$

where t'_{D_s} is the age when the diffusion coefficient becomes stable and βt is a constant (age factor for young concrete). The age t'_{D_s} is normally about 0.5 year for ordinary cement concrete and perhaps a longer time for some concrete containing pozzolanic additions.

The chloride binding is carefully considered in the ClinConc model.

$$\frac{\partial c_b}{\partial c} = k_{OH} \cdot k_{Tb} \cdot \frac{W_{gel}}{1000\varepsilon} \cdot f_b \cdot \beta_b \cdot c^{\beta_b - 1} \qquad (8.7)$$

where c_b is the bound chloride, $\partial cb/\partial c$ is the chloride binding capacity, k_{TD} is the temperature factor for diffusion coefficient, k_{OH} is the factor describing the effect of alkalinity, k_{Tb} is the temperature factor for chloride binding, f_b and β_b are chloride binding constants, W_{gel} is the gel content in kg/m³ concrete, and ε is the water accessible porosity.

In ClinConc, migration coefficient at six months obtained from NT Build 492 is used as input parameter. In order to bridge the gap between the migration coefficient at six months (D_{6m}) and the apparent diffusion coefficient (D_0), Tang introduced a new factor.

$$D_0 = \frac{(0.8a_t^2 - 2a_t + 2.5) \cdot (1 + 0.59K_{b6m}) \cdot k_{TD}}{1 + k_{OH6m} \cdot K_{b6m} \cdot k_{TD} \cdot f_b \cdot \beta_b \cdot \left(\dfrac{c_s}{35.45}\right)^{\beta_b - 1}} \cdot D_{6m} \qquad (8.8)$$

where a_t is the time dependent factor for chloride binding, k_{OH6m} is hydroxide binding coefficient, K_{b6m} is chloride binding factor in non-steady state migration coefficient.

$$\frac{c}{c_s} = 1 - \mathrm{erf}\left(\frac{\chi}{2\sqrt{\frac{k_D \cdot D_0}{1-n}} \cdot \left[\left(1+\frac{t'_{ex}}{t}\right)^{1-n} - \left(\frac{t'_{ex}}{t}\right)^{1-n}\right] \cdot \left(\frac{t'_{6m}}{t}\right)^n \cdot t}\right) \tag{8.9}$$

where n is age factor, and k_D is an extension coefficient.

When the free chloride profile is obtained, the bound and total chlorides can also be obtained.

$$c_b = f_t \cdot k_{OH6m} \cdot K_{b6m} \cdot k_{Tb} \cdot f_b \cdot \beta_b \cdot c^{\beta_b} g/I \tag{8.10}$$

where

$$f_t = a_t \ln\left(\frac{c}{c_s} \cdot t + 0.5\right) + 1 \tag{8.11}$$

From the application of the model, it can be seen that this model can predict the ingress of chloride into concrete quite well.

Kim et al. (2016) carried out a field study to compare the empirical-based model and the ClinConc model. In this study, there were full-scale concrete pier-stems under long-term exposure to a marine environment, with work focusing on XS2 (below mid-tide level) in which the concrete is regarded as fully saturated and XS3 (tidal, splash, and spray) in which the concrete is in an unsaturated condition. These exposures represent zones where concrete structures are most susceptible to ionic ingress and deterioration. Chloride profiles and chloride transport behavior are studied using both an empirical model (erfc function) and a physical model (ClinConc). The time dependency of surface chloride concentration (C_s) and apparent diffusivity (D_a) were established for the empirical model whereas, in the ClinConc model (originally based on saturated concrete), two new environmental factors were introduced for the XS3 environmental exposure zone. Although the XS3 is considered as one environmental exposure zone according to BS EN 206-1:2013, the work has highlighted that even within this zone, significant changes in chloride ingress are evident.

(2) Model by Samson (Stadium, 1999)
  • Equations describing the transport process
    – The mass balance equation
    – The extended Nernst–Planck equation

- The Poisson equation
- Chloride binding–chemical equilibrium
- The effect of temperature
- Inputs
  - Migration coefficients of various ions. Chloride migration coefficient is determined by measuring the electrical current passing through the sample over a period. The results are then analyzed by the set of extended Nernst–Planck equation. An algorithm is used to determine for each experiment the tortuosity factor that allow to best reproduce the current curve measured experimentally. Then, the migration coefficient can be obtained. The migration coefficients of other species are computed by assuming that the ratio of its diffusion coefficient to chloride diffusion coefficient at infinite dilution equals to the ratio of its migration coefficient to determine chloride migration coefficient.
  - The material conditions (the chemical composition of pore solution, the initial composition of material, the total and capillary porosity).
  - The environmental conditions (ionic concentrations and types of exposure solution and temperature).
- Outputs
  - The ionic profiles (free and total Cl^-, K^+, Na^+, Ca^{2+}, OH^-, SO_4^{2-})
  - The solid phase distribution
  - The flux of each ion inside and outside concrete
  - The potential distribution across the concrete (in the case of migration)

Samson and Marchand have been continuously improving their model during the past ten years (Samson et al. 1999a, 1999b, 1999c, 1999d, 2000, 2003; Samson and Marchand 2007; Marchand 2001). Their research results were published in various journals, from chemistry and computation to building materials. Their model can be used in the case of diffusion, electrical migration, and even capillary suction, and has become one of the most sophisticated models. Coupled extended Nernst–Planck equation and Poisson equation were numerically solved by the finite element method. According to Samson et al. (1999a), the Newton–Raphson method appeared to be more suitable than the Picard iteration method and was therefore adopted for analysis. The chemical equilibrium is verified at each point. The effect of temperature on the chemical equilibrium is taken into account. As many chemical and physical processes as possible are taken into account in this model. However, many data are based on electrolytes. Whether it applies to cement-based materials still needs further research. Some of the input data, like the chemical composition of pore solution, are not easy to get, and reaction enthalpy of some cement hydration products, such as C-S-H and Friedel's salt, are not available. The complexity of the model makes it impractical for engineers and not easy to understand.

(3) Model by Truc (Msdiff, 2000)
- Equations describe the transport process
  - The mass balance equation
  - The Nernst–Planck equation
  - Eletroneutrality equation
  - Chloride binding relationship—Freundlich isotherm.
- Inputs
  - Migration coefficient of various ions. Chloride effective migration coefficient is determined by a so called "upstream method," which is a modified version of NT Build 355. The migration coefficients of other species are obtained by the same method Samson adopted.
  - Materials conditions (density of concrete, porosity, pore solution chemistry, chloride binding isotherm).
  - Environmental conditions (ionic concentrations and type of exposure solution).
- Outputs
  - The ionic profiles (free and total Cl-, K+, Na+, OH-)
  - The evolution of the porosity and dry density
  - The flux of each ion inside and outside concrete
  - The potential distribution across the concrete (in the case of migration)

Truc used an electroneutrality equation, instead of the Poisson equation, which is regarded to be a good approximation of Poisson equation. Then, the finite difference method was applied to solve the set of partial differential equations. Like Samson's model, Truc's model can be used in the case of diffusion or migration. Pore solution chemistry and the effective diffusion coefficient are the most important input parameters. However, pore solution chemistry is not easy to obtain, and the accuracy of effective diffusion coefficient obtained by upstream method still needs further investigation. Based on Truc's model, Khitab et al. (2005) extended this model to a more practical version.

Since chloride is a charged particle, the multi species model is discussed in detail in this section based on works by Truc (2000) and Yuan (2009). The ionic flux through a saturated porous medium is often given by the extended Nernst–Planck equation:

$$J = -D\left[\frac{\partial c}{\partial x} + \frac{c}{\gamma}\frac{\partial \gamma}{\partial x} + zc\frac{F}{RT}\frac{\partial E(x,t)}{\partial x} - cV(x)\right] \qquad (8.12)$$

where D is the effective diffusion coefficient, c_i is the ionic concentration in pore solution, γ is the chemical activity coefficient, E is the electrical potential, F is the Faraday constant, R is the universal gas constant, and z is the valence.

Each term on the right-hand side of Equation 8.12 corresponds to different mechanisms. The first term, often called the diffusion term or Fick's first law, describes the movement of ions under the effect of a concentration gradient.

The second term represents the effect of chemical activity coefficient. In an infinite dilution, the chemical activity coefficient equals 1; while it is lower than 1 in a normal solution. Classical electrochemical models like the Debye–Hückel or extended Debye–Hückel relationships are valid for weak electrolytes for which the ionic strength is on the order of 100 mmol/l. According to Samson et al. (1999b), the Davies correction can be used to describe the behavior of more concentrated solutions, i.e., with ionic strengths up to 300 mmol/L. Pore solutions are more in the range of 300–900 mmol/l. The influence of chemical activity coefficient on the transport process has been widely studied. Truc (2000) used a model based on Pitzer equations and calculated the fluxes of ionic species and the potential created by the movement of different species across the specimen under both steady- and non-steady-states. The calculated fluxes are shown in Table 8.1. Results showed that the difference between the flux of ideal and non-ideal electrolytes is very small. The difference on electrical potential across the specimen is also very small. Samson et al. (1999b) used a modified version of Davies law, which is able to calculate the chemical activity coefficient for an electrolyte with high ionic strength. The numerical results presented by Samson also showed that the chemical activity gradient has only a minor influence on the transport of ions. Tang also reached the same conclusions (Tang 1999). Most of the published results show that the influence of chemical activity gradient on the transport of ions is negligible. Therefore, the second term in Equation 8.12 is omitted.

The third term in Equation 8.12 is the effect of electrical field, which may be the combination of membrane potential and external electrical field.

Table 8.1 Influence of Activity of the Electrolyte on the Values of the Flux (for $x = 1$ mm) [10^{-8}mol/(m^2/s)]

	No Activity in ss	Activity in ss	No Activity in nss	Activity in nss
Na^+	0.0079	0.0078	0.0122	0.0123
K^+	-0.0128	-0.0128	-0.0293	-0.0291
Cl^-	0.1140	0.1127	0.5240	0.5251
OH^-	-0.2475	-0.2481	-1.3714	-1.3688

Truc (2000).

ss: steady-state; nss: non-steady-state

When the external electrical field is large enough, the membrane potential as well as the diffusion term could be negligible. This is the case in NT Build 355.

Zhang and Buenfeld (1997) measured the membrane potential of 20–45mv across ordinary Portland cement (OPC) mortar specimens housed in chloride diffusion cells. The electrical field of this magnitude is believed to have a significant influence on the ions' diffusion in cementitious materials. Truc (2000b) used a model called MSdiff to calculate the membrane potential across cement paste and reached a value very close to Zhang's results. Thus, the membrane potential term cannot be neglected in the case of diffusion.

The fourth one is the advection term. No pressure gradient is considered. Accordingly, the advection term in Equation 8.12 is dropped.

Thus, Equation 8.12 becomes:

$$J = -D\left[\frac{\partial c}{\partial x} + zc\frac{F}{RT}\frac{\partial E(x,t)}{\partial x}\right] \tag{8.13}$$

Continuity equation is given as follows:

$$\frac{\partial J}{\partial x} = -\frac{\partial c_t}{\partial t} = -\frac{\partial(pc + (1-p)\rho c_b)}{\partial t} \tag{8.14}$$

Introduce Equation 8.13 into Equation 8.14, and then the continuity equation turns into:

$$p\frac{\partial c}{\partial t} + (1-p)\rho_{dry}\frac{\partial c_b}{\partial t} = \frac{\partial}{\partial x}\left(D\left[\frac{\partial c}{\partial x} + Zc\frac{F}{RT}\frac{\partial E(x,t)}{\partial x}\right]\right) \tag{8.15}$$

where p is the water accessible porosity, c is the free chloride concentration in concrete (mol/l), ρ_{dry} is the dry density of concrete (kg/m³), and c_b is the number of mole of chloride bound by the dry concrete (mol/kg).

The first term on the left-hand side represents the free chloride term, and the second term is the bound chloride term. Chloride binding is concentration dependent; thus,

$$\frac{\partial c_b}{\partial t} = \frac{\partial c_b}{\partial c}\frac{\partial c}{\partial t} \tag{8.16}$$

Then,

$$\frac{\partial c}{\partial t} = \left(p + (1-p)\rho\frac{\partial c_b}{\partial c}\right)^{-1}\frac{\partial}{\partial x}\left(D\left[\frac{\partial c}{\partial x} + Zc\frac{F}{RT}\frac{\partial E(x,t)}{\partial x}\right]\right) \tag{8.17}$$

To evaluate the electrical potential E(x,t), Poisson's equation has to be used to relate electrical potential and electrical charge $\sum_i c_i Z_i$ in solution. Samson et al. (1999a, 1999d, 2003, 2007) solved coupled Poisson's equation and Nernst–Planck equation with finite element method. Truc (2000) used another strategy, which is used in this study.

In the pore solution, the different ionic species must respect the electroneutrality condition, which is a very good approximation of Poisson's equation:

$$\sum_i c_i Z_i = 0 \tag{8.18}$$

where the subscript i represents the type of ions. During a diffusion process, the current should be equal to 0

$$F \sum_i z_i J_i = 0 \tag{8.19}$$

Then the electrical field created by the movement of different species can be obtained from Equations 8.19 and 8.14.

$$\frac{\partial E}{\partial x} = -\frac{RT}{F} \frac{\sum_i z_i D_i \left(\frac{\partial c_i}{\partial x} \right)}{\sum_i Z_i^2 D_i c_i} \tag{8.20}$$

A set of equations for the simulation of multi species transport in cementitious materials has been established. In order to simulate the transport process, Equations 8.17 and 8.20 have to be solved uni-dimensionally. The first step is to determine the types of species of interest and their diffusion coefficients.

It is well known that pore solution of cementitious materials is concentrated, mainly consisting of Na^+, K^+, Ca^{2+}, SO_4^{2-}, OH^-, Cl^-. Wiens et al. (1995) studied the chemistry of pore solution of various cement paste. Results showed that the concentrations of K^+, Na^+, OH^- are quite high, while the concentrations of Ca^{2+}, SO_4^{2-}, Cl^- are relatively low. When exposed to chloride environments, Cl^- is the one of interest. Therefore, only four types of species, i.e., K^+, Na^+, OH^-, Cl^- are taken into consideration.

The diffusion coefficient of chloride ion can be experimentally determined. The diffusion coefficients of other species are calculated by assuming that the ratios between diffusion coefficients in infinite diluted solution are equal to that in cementitious materials; hence,

$$\left(\frac{D_{Cl}}{D_i} \right)_{cem} = \left(\frac{D_{Cl}}{D_i} \right)_{inf} \tag{8.21}$$

where the subscript cem represents in cement materials, and subscript inf corresponds to the infinite diluted solution. Diffusion coefficients of various species in infinite dilution are shown in Table 8.2.

Table 8.2 Diffusion Coefficients of Various Species in Infinite Dilution ($10^{-9} m^2/s$)

Species	Diffusion Coefficient
Na^+	1.334
K^+	1.957
OH^-	5.273
Cl^-	2.032
Ca^{2+}	0.792
SO_4^{2-}	1.065

Samson et al. (2003).

Nevertheless, Equation 8.21 does not hold in reality. It is shown that the $D_{Anions}/D_{Cations}$ ratio in mortar is larger in pore water than at infinite dilution (Goto and Roy 1981; Truc 2000),

$$\left(\frac{D_{Cl}}{D_{cation}}\right)_{cem} > \left(\frac{D_{Cl}}{D_{cation}}\right)_{inf} \tag{8.22}$$

Introduce a factor of r, then,

$$\left(D_{cation}\right)_{cem} = r\left(\frac{D_{cation}}{D_{Cl}}\right)_{inf}\left(D_{Cl}\right)_{cem} \tag{8.23}$$

Truc (2000b) conducted numerical analysis by using different ratios lower than 20. Results showed that the factor of r has not a significant effect on the concentration profiles. It just slightly changed the shape of the curve by more or less slowing down the chloride penetration. Therefore, Equation 8.21 is used in the model.

To solve Equation 8.17, the binding of various species to cement hydration products have to be known. The binding of cations is often neglected. Only the binding of chloride ions is taken into account, since it is the species of interest. Nevertheless, chloride binding is a very complicated phenomenon and influenced by many other species. For example, hydroxyl ions have a significant negative influence on chloride binding. Sulfate ions, as well as the type of cations, also influence chloride binding. This has been reviewed in depth in Chapter 4. Tang (1996) and Truc (2000b) implemented the effect of hydroxyl ions on chloride binding in their models, and reached quite good results. Freundlich binding isotherm is applied in their models, which is determined by equilibrium method. Freundlich binding isotherm is also adopted in the model, which is determined from diffusion/migration tests. This has been discussed in Yuan et al. (2013), Figure 8.7.

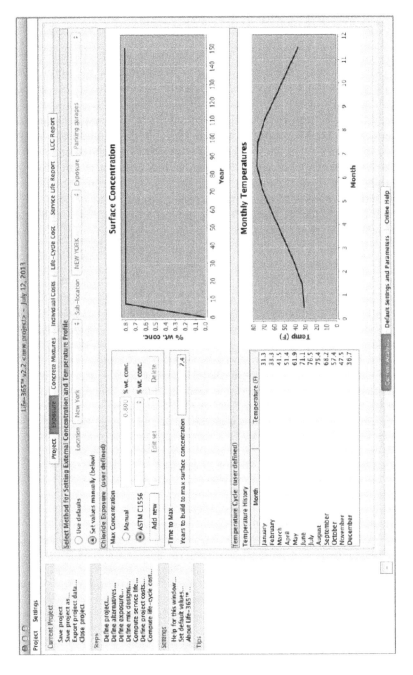

Figure 8.7 Evolution of surface chloride concentration.

$$c_b = ac^\beta \qquad\qquad (8.24)$$

where c_b is expressed in mol/kg of concrete, c is expressed in mol/m³ of pore solution.

The last question to solve the set of equations is which chloride coefficient should be used as the input parameter for the service life prediction model. Tang used the testing results determined according to NT Build 492 which is then converted to effective diffusion coefficient in his model. Truc (2000b) and Khitab et al. (2005) adopted the testing results obtained from upstream method. Samson et al. (2003) obtained migration coefficients by regularly measuring current passing through the sample over a period. The time dependence of the diffusion coefficient might also be taken into account in all the models. Nevertheless, the diffusion coefficient is considered as constant along the penetration direction. In other words, the diffusion coefficient is not depth dependent. If looking at a chloride profile, one will find that the diffusion coefficient is depth dependent. There are two reasons for this. First, electrochemical reasons; chloride diffusion coefficient is concentration dependent (Tang 1999, Truc 2000b, Zhang 1997). The chloride diffusion coefficient at point A differs from that of point B, since their concentrations are different, as illustrated in Figure 8.8. Secondly, microstructural reasons; MIP results showed that bound chloride changes pore structure of concrete significantly (Yuan et al. 2011). Because the chloride concentration of point A is higher than that of point B, point A may bind more chloride than point B. therefore, the pore structure of point A is changed more significantly than that of point B. Consequently, it affects diffusion coefficient in the same fashion. Thus, a varying diffusion

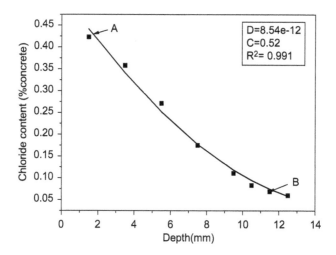

Figure 8.8 Chloride concentrations at different depths.

coefficient (along the penetration direction), instead of a constant diffusion coefficient, should be used in the service life prediction model.

Steady-state diffusion coefficient should be used in the prediction model. Since it takes very long time to conduct steady-state diffusion test, migration test is adopted instead. Among the two existing methods for steady-state migration coefficient, the results from upstream method were higher than that from downstream method. This is because there is a theoretical error in upstream method. Thus, downstream method, i.e., NT Build 355, is adopted. On the other hand, steady-state migration coefficient is generally higher than steady-state diffusion coefficient (Arsenault et al. 1995). This is solved by introducing a factor of 1.6 (Yuan et al. 2011). In combination with the concentration dependency of migration coefficients, the input parameter of diffusion coefficient can be expressed as follows:

$$
\begin{cases}
D = D_{0.01M}/1.6 \ldots\ldots\ldots (c < 0.01 \text{ mol/l}) \\
D = D_{1M} \times c^{-0.7}/1.6 \ldots\ldots (0.01 \text{ mol/l} \leq c \leq 1 \text{ mol/l}) \\
D = D_{1M}/1.6 \ldots\ldots\ldots (1 \text{ mol/l} < c)
\end{cases}
\tag{8.25}
$$

where D_{1M} is the steady-state migration coefficient determined from NT Build 355 at 1 mol/l NaCl solution. $D_{0.01M}$ is steady-state migration coefficient at 0.01 mol/l. By doing this, the concentration dependency of the chloride diffusion coefficient, which includes electrochemical effects and microstructural effects, is taken into account.

The set of partial differential equations describing chloride transport in concrete is very complicated and has to be solved numerically. Finite difference method, which is very useful in solving partial differential equations, was adopted. A commercial software, Matlab 7.0, was used to obtain the numerical results.

As mentioned earlier, four types of species, i.e., K^+, Na^+, OH^-, Cl^-, are taken into consideration. Equation 8.17 can be written as:

$$
\frac{\partial E}{\partial x} = -\frac{RT}{F} \frac{D_{Na}\frac{\partial c_{Na}}{\partial x} + D_K\frac{\partial c_k}{\partial x} - D_{Cl}\frac{\partial c_{Cl}}{\partial x} - D_{OH}\frac{\partial c_{OH}}{\partial x}}{D_{Na} \cdot c_{Na} + D_K \cdot c_K + D_{Cl} \cdot c_{Cl} + D_{OH} \cdot c_{OH}}
\tag{8.26}
$$

At every time step, Equation 8.26, i.e., the electrical field equation, is solved first with a centered explicit scheme. The resulting electrical potential is then introduced into the continuity equation, i.e., Equation 8.18. K^+, Na^+, Cl^- are obtained from the continuity equation, and OH^- is obtained from the electroneutrality equation, then,

$$\begin{cases} \dfrac{\partial c_{Na}}{\partial t} = \dfrac{\partial}{\partial x}\left(D_{Na}\left[\dfrac{\partial c_{Na}}{\partial x} + c_{Na}\dfrac{F}{RT}\dfrac{\partial E(x,t)}{\partial x} \right] \right) \\[3mm] \dfrac{\partial c_{K}}{\partial t} = \dfrac{\partial}{\partial x}\left(D_{K}\left[\dfrac{\partial c_{K}}{\partial x} + c_{K}\dfrac{F}{RT}\dfrac{\partial E(x,t)}{\partial x} \right] \right) \\[3mm] \dfrac{\partial c_{Cl}}{\partial t} = \left(p + (1-p)\rho_{dry}\dfrac{\partial c_{b,Cl}}{\partial c_{Cl}} \right)^{-1}\dfrac{\partial}{\partial x}\left(D_{Cl}\left[\dfrac{\partial c_{Cl}}{\partial x} - c_{Cl}\dfrac{F}{RT}\dfrac{\partial E(x,t)}{\partial x} \right] \right) \\[3mm] c_{OH} = c_{Na} + c_{k} - c_{Cl} \end{cases} \qquad (8.27)$$

The above equations were written in their own finite difference schemes under Matlab 7.0. With initial boundary conditions and other input parameters, the ionic profiles can be computed. Total and free chloride profiles are the ones of interest, which can be given by this model. Hydroxyl ion concentration is very important to the corrosion initiation of steel. The hydroxyl ion profile may also be obtained.

In order to solve the equations numerically by finite difference method, the initial and boundary conditions have to be known. With respect to the boundary conditions, researchers often take the chemical composition and ionic concentrations of the exposure solution as boundary conditions in laboratory test, which is the case in this study. This implicitly means that the surface free chloride concentration is equal to that of the exposure solution. However, the previous study showed that in some cases the surface free chloride concentration might be lower than that of exposure solution after immersing in 165 g/l NaCl solution for 42 days. It is reasonable to assume that the surface free chloride concentration will be equal to that of exposure solution at some time. The time to reach the equilibrium depends on the w/b ratio, supplementary cementing materials etc. This is not well understood. In reality, however, the boundary conditions are much more difficult to obtain. As mentioned earlier, surface chloride concentration changes with time, and will reach a constant value after a certain exposure time. For simplicity, the chemistry of exposure solution was taken as the boundary conditions.

Concerning the initial conditions, the chemistry of pore solution has been discussed earlier. It is not easy to obtain pore solution from concrete with w/b ratio less than 0.5. Some researchers (Reardon 1992; Schmidt and Rostasy 1993) developed software to estimate the chemical composition of pore solution, but the uncertainties of the results are not easy to evaluate. In this research, the pore solution chemistry was roughly estimated from the published data (Schmidt and Rostasy 1993; Wiens et al. 1995; Samson et al. 2003).

The other input parameters needed for the simulation include:

- Binding constants
- Effective diffusion coefficients of various ions
- Porosity and dry density of concrete

The binding constants can be experimentally determined, as described in Chapter 4. Note that the unit for free chloride is mol/m^3 pore solution, and bound chloride is expressed in mol/kg dry concrete. The effective diffusion coefficients of chloride ions were determined from NT Build 355 at 1 mol/l NaCl, and then introduced into Equation 8.27 in which the concentration dependency of migration coefficient, and difference between migration and diffusion tests were taken into consideration. The effective diffusion coefficients of other ions are obtained by assuming that the ratios between diffusion coefficients in infinite diluted solution are equal to that in cementitious materials, as shown in Equation 8.21. The porosity and dry density of concrete were experimentally measured.

The experimental data of blank concrete with the w/b ratio of 0.48 was validated with the prediction model (Yuan 2009). All the necessary input data are given in Table 8.3.

Figure 8.9 gives the comparison of the experimental data with the results predicted by models implemented with varying and fixed effective diffusion

Table 8.3 Input Data for the Prediction Model

Mix		B48	
Binding constant	α	0.32×10^{-3}	
	β	0.68	
D_{ssd} ($\times 10^{-12}$ m^2/s) at 1 mol/l	Ions	D_{nssm}	$D_{nssm}/1.6$
	K$^+$	0.48	0.3
	Na$^+$	0.33	0.206
	Cl$^-$	0.5	0.313
	OH$^-$	1.30	0.813
Porosity (%)		14.1	
Dry density(kg/m^3)		2286	
Pore solution chemistry (mmol/l)	K$^+$	320	
	Na$^+$	100	
	Cl$^-$	0	
	OH$^-$	420	
Exposure solution chemistry (mmol/l)	K$^+$	0	
	Na$^+$	2800	
	Cl$^-$	2800	
	OH$^-$	0	

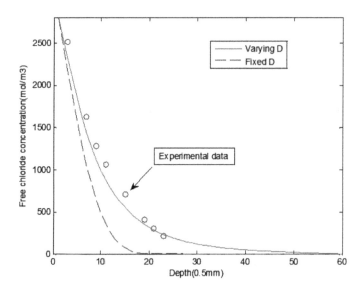

Figure 8.9 Comparison of the experimental data with the simulation results predicted by models implemented with varying and fixed effective diffusion coefficient respectively. (From Yuan 2009.)

coefficients respectively. It can be seen that the experimental data are in perfect agreement with the results predicted by the model implemented with varying effective diffusion coefficient (i.e., Equation 8.25). In contrast, the results predicted by the model using the fixed effective diffusion coefficient greatly underestimate the chloride profile.

As stated earlier, diffusion coefficients determined from electrically accelerated tests are generally higher than diffusion coefficients determined from diffusion tests. A factor of 1.6 thus was introduced. When the factor of 1.6 between diffusion tests and migration tests is not taken into account, the model implemented with the varying D_{nssd} overestimates the chloride front, as shown in Figure 8.10, while the model implemented with the fixed D_{nssd} still underestimates the chloride front.

Figure 8.11 gives the comparison of experimental data on free and total chloride profiles in the unit of % by the mass of dry concrete with the free and total chloride profiles predicted by the model. It can be seen that this model can predict both free and total chloride profiles quite accurately.

Truc (2000b) used the multi-species model to predict total chloride profile of concrete after immersion of 10 months, and found that the model underestimated total chloride near to the exposure surface, and overestimated the chloride front. Truc ascribed this to the chloride isotherm used in this model which is determined by equilibrium method: first, the testing period for equilibrium method may not long enough to reach equilibrium; second, the effect of hydroxyl ions on the chloride binding is not taken into

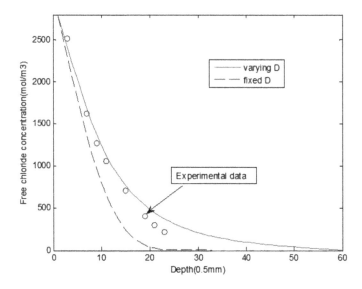

Figure 8.10 Comparison of the experimental data with the simulation results predicted by models implemented with varying and fixed effective diffusion coefficient without considering the factor of 1.6 between migration and diffusion tests. (From Yuan 2009.)

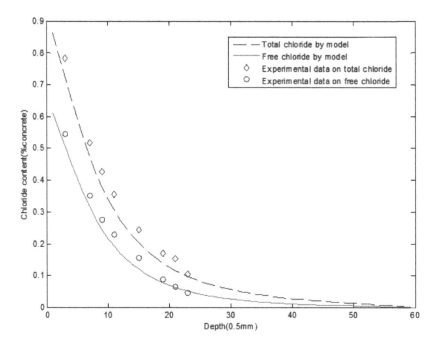

Figure 8.11 Comparison of the experimental data on total and free chloride profiles with the simulation results predicted by the model. (From Yuan 2009.)

account in the equilibrium method. To compensate the second drawback, Tang (1996) proposed a hydroxyl dependent factor:

$$f_{OH} = \frac{B_{Cl}}{B_{Cl[OH]_{ini}}} = e^{0.59\left(1-\frac{[OH]}{[OH]_{ini}}\right)} \tag{8.28}$$

where f_{OH} is the hydroxyl dependent factor, $B_{Cl[OH]ini}$ is the bound chloride obtained by equilibrium method, and B_{Cl} is the corrected bound chloride. [OH] is the hydroxyl ion concentration in pore solution, and $[OH]_{ini}$ is the hydroxyl ion concentration in the exposure solution in equilibrium method.

This phenomenon is not observed in Figure 8.11; experimental total chloride profile and predicted profile are in perfect agreement. There is no need to use the correction factor for pH value. This is because the chloride binding isotherm implemented in the model was directly determined from the specimens after diffusion or migration tests (Yuan et al. 2013). Actually, the effect of hydroxyl ions on binding was implicitly taken into account in the method. This is an advantage. Nevertheless, the prediction was conducted on short-term experiments in this study. The applicability of this method for long-term prediction is still unknown and needs further investigation.

It is worth it to mention that the corrosion initiation of steel bar is not governed by the chloride concentration but by the ratio of Cl⁻/OH⁻ of pore solution at the surface of steel bar. When this ratio exceeds a threshold, the corrosion of steel will be initiated. Due to difficulties in the measurement of the ratio of Cl⁻/OH⁻ and dependency of corrosion initiation on many factors, published data showed a large variability in the threshold. It may vary from 0.6 to 3.0 (Hussain et al. 1995). Therefore, to know the hydroxyl ion concentration at the surface of steel bar is also very important. This model can predict hydroxyl ions profile, as shown in Figure 8.12. It can be seen that the more chloride ions penetrate into concrete, the more hydroxyl ions move into solution. This implicitly means that the hydroxyl ions' concentration in the exposure solution should be increased. Nevertheless, it can be noticed that the hydroxyl ions at $x = 0$ equals to zero. This is because the boundary conditions, in which chloride concentration is equal to zero, were taken as constants.

In summary, the multi species model discussed in this section, in which the varying diffusion coefficient and the binding isotherm determined from diffusion tests were used, can predict the total and free chloride profile in the diffusion test accurately. In many other models, the time dependency of the diffusion coefficient was taken into account. It also can be easily included in the model proposed in this study.

However, the model proposed in this study is still far from practical. The real conditions are much more complicated than the laboratory

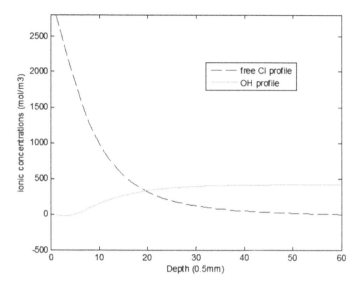

Figure 8.12 Free chloride and hydroxyl ions profile predicted by the model. (From Yuan 2009.)

conditions. Prediction of chloride-induced corrosion in the field still faces many challenges:

- Chloride transport in concrete is often governed by different mechanisms, not only diffusion. Coupling the multi species with other transport mechanisms is necessary. Samson has made some progress in this respect (Samson and Marchand 2007).
- Chloride-induced corrosion is often associated with other detrimental effects, like freezing–thawing, sulfate attack, carbonation, etc.
- Boundary and initial conditions are necessary for numerical simulation. The boundary conditions vary with time, vary from area to area, and from structure to structure. For the bridge deck subjected to de-icing salt, the de-icing salt is not continuously used. It depends on the weather, maybe a few days or a few months. For the structure near the sea, the splash zone, immersion zone, and the zone exposed to the air have different boundary conditions. How to define and quantify the boundary conditions is a challenge. The initial ionic concentrations of pore solution of concrete with low w/b ratio are also difficult to determine.
- The decomposition and leakage of cement hydration products are not taken into account in this model. This affects the concentrations of pore solution and chloride binding greatly.

(4) Model by Tran et al. (2018)
- Equations describe the transport process
  - The mass balance equation
  - Fick's second law
  - Thermodynamic equilibrium/kinetic chemical
  - Eletroneutrality equation
  - Surface complexation model
- Inputs
  - Mineralogical compositions of hydrated cement
  - Reaction rate and surface complexes parameters
  - Initial mineralogical composition, pore solution of concrete and boundary conditions;
  - Thermodynamic database
  - Migration coefficient of various ions
  - Materials conditions (density of concrete, porosity, pore solution chemistry, chloride binding isotherm)
- Outputs
  - The ionic profiles (free and total Cl⁻, and other species)
  - The evolution of the porosity and dry density
  - Chloride binding isotherm

In this model, a new physically and chemically based model taking into account thermodynamic equilibrium, kinetics, and surface complexation is proposed to predict the ingress of chloride ions into saturated concretes.

There are also many other models (Johannesson et al. 2007; Nguyen et al. 2006; Fenaux et al. 2019) which simulate the chloride transport in concrete. These models will not be introduced in detail here. Fenaux et al. (2019) proposed a model which couples the transport equations of ionic species in concrete with the Pitzer model and allows simulation of the transport of the ionic species present in the pore solution and considers diffusion, migration, and chemical activity. The outputs of the model are the concentration profiles of all the ionic species present in the pore solution (total, free, and bound), pH of the pore solution, electric field, chemical activity, and activity coefficient. It was found that the effect of chemical activity on chloride and sodium transport appeared to be negligible, whereas the other species seemed to be more affected. If the chloride ions are the ionic species of interest, chemical activity can be ignored and the coupling of the transport equations to the Pitzer model can be omitted. The electric field significantly affects penetration of ionic species.

8.2.3 Probabilistic Model

Due to the stochastic nature of the parameters in the model, durability design concepts have the trend that moving from a deterministic method to a probabilistic approach (Gjørv 2013; Altmann and Mechtcherine 2013; Samindi et al. 2015).

From a probabilistic point of view, the basic structural reliability problem considers only one load effect S resisted by one resistance R. Both S and R are described by known probability density functions. The failure probability can be expressed as:

$$p_f = P(R - S \le 0) \tag{8.29}$$

The service life of concrete structures can be expressed in Figure 8.13. The error function solution to Fick's second law is often used as the basic equation (Lindvall 1999, 2002; Kirkpatrick et al. 2002; Williamson et al. 2008). Some factors were introduced to improve this model, such as time-dependent diffusion coefficient, time-dependent surface chloride concentration, curing factor, temperature factor, environmental factor, etc. The normal form of the equation is as follows:

$$C_x = C_s \left[1 - \mathrm{erf} \frac{x}{2\sqrt{D_0 k_1 k_2 \left(\frac{t_0}{t}\right)^m t}} \right] \tag{8.30}$$

where C_x is the chloride content as a function of the distance to the surface, C_s is the chloride content at the surface, D_0 is the reference diffusion coefficient, m is the time-dependent factor, k_1 and k_2 are the curing factor and environmental factor, t_0 is the reference time, t is the exposure time.

Different probability density functions are used to describe the variables presented in Equation 8.30. In the DuraCrete project (Lindvall 1999, 2002)

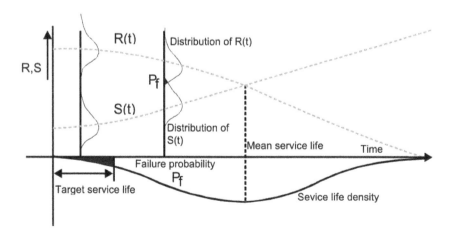

Figure 8.13 Failure probability and target service life.

the chloride corrosion initiation concentration and diffusion coefficient are considered normally distributed, and cover depth and surface chloride concentration are lognormal distributions. A diffusion coefficient geographic location parameter is considered a beta distribution, and a time diffusion coefficient reduction factor is considered to be a gamma distribution. Monte Carlo simulation techniques are often used to solve the probabilistic problem.

In principle a probabilistic method has been considered as a suitable approach for service life of reinforced concrete structures because of the stochastic nature of the model parameters in reality. However, the statistical values of the model parameters are far from sound. Gulikers (2007) has shown that the outputs are very sensitive to the input values used. Therefore, a comprehensive data base of the model parameters should be established for the service life prediction model.

In the final report of the Duracrete project (1998), the Load and Resistance Factor Design (LRFD) method was recommended. This method is based on: (1) design equations; (2) characteristic values of load and resistance variables, defined as a fractal of the probability distribution function of the given parameter; (3) partial factors for the load and resistance variables.

The four variables are considered in this model.

- Material variables
- Environment variables
- Execution variables
- Variables depending on both the environment and the materials

The design equations are:

$$g = c_{cr}^d - c^d(x,t) = c_{cr}^d - c_{s,cl}^d \left[1 - \mathrm{erf} \left(\frac{x^d}{2\sqrt{\dfrac{t}{R_{cl}^d(t)}}} \right) \right] \tag{8.31}$$

$$c_{cr}^d = c_{cr}^c \cdot \frac{1}{\gamma_{c_{cr}}} \tag{8.32}$$

$$c_{s,cl}^d = A_{c_{s,cl}} \cdot (w/b) \cdot \gamma_{c_{s,cl}} \tag{8.33}$$

$$x^d = x^c - \cdot x \tag{8.34}$$

$$R_{cl}^d(t) = \frac{R_{cl,0}^c}{k_{e,cl}^c \cdot k_{c,cl}^c \cdot \left(\dfrac{t_0}{t}\right)^{n_{Cl}^c} \cdot \gamma_{R_{cl}}} \tag{8.35}$$

where $\gamma_{c_{cr}}$ is the partial factor of the critical chloride concentration; $A_{c_{s,cl}}$ is a regression parameter describing the relationship between the surface chloride concentration and w/c ratio, and $\gamma_{c_{s,cl}}$ is the partial factor for the surface concentration. Δx is the margin for the cover thickness. $R_{cl,0}$ is the resistance with respect to the chloride ingress determined on compliance tests; $k_{c,cl}$ is curing factor; $k_{c,e}$ is environment factor; t_0 is the age of the concrete when the compliance test is performed. n_{cl} is age factor. γ_{Rcl} is partial factor for the resistance with respect to chloride ingress.

The characteristic values and partial factors were provided in Duracrete (1998); thus, the time to corrosion initiation can be calculated accordingly.

8.3 MODELING CHLORIDE TRANSPORT IN UNSATURATED CONCRETE

It is obvious that most concretes in service are unsaturated. It is of more practical significance to predict chloride transport in unsaturated concrete. In comparison with chloride transport in saturated concrete, however, the degree of water saturated has a great influence on the transport of chloride, and remarkably increases the complexity of the transport process. Substantial experimental works and simulation works are needed to better understand this transport process, as reviewed by Zhang et al. (2014).

Because of its crucial role in the service life prediction and durability assessment of reinforced concrete structures in service, an increasing number of experimental and modeling studies related to transport properties in unsaturated concrete have been and still are performed. In this section, the model for chloride transport in unsaturated concrete are discussed in two categories, i.e., deterministic model and probabilistic model.

8.3.1 Deterministic Model

In unsaturated concrete, chloride may transport into concrete in several different mechanisms. Due to its complexity, it is difficult or even impossible to precisely quantify the effects of these combined processes on the chloride transport. In order to predict chloride transport in unsaturated concrete more precisely, many attempts have been devoted to develop physical models based on the physical–chemical mechanisms.

Saetta et al. (1993) and Ababneh et al. (2003) developed a model coupled chloride transport and moisture transfer based on Fick's second law. Martys (1999) developed a multicomponent lattice Boltzmann model to predict the combined moisture and chloride transport. In order to overcome the drawbacks of Martys's model, Zhang et al. (2012) developed a modified Shan-Chen multiphase lattice Boltzmann model by incorporating different kinds of equations of state (EOS) and introducing a parameter named the virtual density of wall into the original Shan-Chen model. Petcherdchoo (2018) developed a set of new closed-form solutions for predicting the transport of chloride ions in unsaturated concrete. The proposed solutions can be used in the situation that possesses both convection and diffusion. Based on extended Nernst–Planck equations, Samson and Marchand (2007) and Nguyen and Amiri (2014, 2016) developed a complex physical model where the diffusion, migration, and convection mechanisms are considered. In this model, the homogenization technique is applied to a microscopic equation for both diffusive and convective transport.

$$\frac{\partial\left(wk \mp C_i\right)}{\partial t}+\left(1-\varepsilon_0\right)\frac{\partial\left(K \mp C_{ib}\right)}{\partial t}$$

$$-\mathrm{div}\left(\underbrace{wD_iK \mp \mathrm{grad}C_i}_{\text{diffusion}}+\underbrace{\frac{K \mp D_iFz_i}{RT}C_iw\mathrm{grad}\Psi}_{\text{migration}}-\underbrace{K \mp wC_iU}_{\text{convection}}\right)=0$$

(8.36)

where ε_0 is the porosity of the material, w is water content. C_i is the concentration of chemical specie i contained in the pore solution. C_{ib} is the concentration of specie i bound to cement matrix of the concrete. ψ is the electrical potential in the pore solution. D_i is the diffusion coefficient of the specie i and z_i is its valence. F is the Faraday's constant, R is the gas constant, and T is the thermodynamic temperature (293K).

$$\mathrm{grad}\Psi = -\frac{\displaystyle\sum_{i=1}^{n}D_iK \mp C_iz_iw\mathrm{grad}C_i - \sum_{i=1}^{n}z_iwUK \mp C_i}{\dfrac{F}{RT}\displaystyle\sum_{i=1}^{n}D_iK \mp C_{ib}z_i^2w\nabla C_{ib}}$$

(8.37)

$$K_+ = \frac{A\left(\kappa d\right)^2}{1-AB\kappa d + A\left(\kappa d\right)^2}+\frac{1+Z}{2}\frac{2q}{FdC_{pc}}$$

(8.38)

$$K_- = K_+ - \frac{2q}{FdC_{pc}}$$

(8.39)

A and B are two parameters which depend on the zeta potential and the pore diameter d, the concentration of the bulk Cpc, the surface density charge q and on Debye constant κ, Z is the sign of the electrical double layer (EDL) (Z = 1, EDL positive; Z = -1, EDL negative).

$$D_w = -\frac{K_lK_{rl}}{\mu_l}\frac{\partial P_c}{\partial \chi}+\frac{D_{va}}{\rho_w}\frac{\partial \rho_v}{\partial w}$$

(8.40)

$$D_{va} = 2.17 \times 10^{-5} \left(\frac{T}{T_{ref}} \right)^{1.88} * f_{res}$$

where K_1, K_{r1}, μ_l are the intrinsic permeability, relative permeability, and the density of water, respectively. T and T_{ref} are the temperature of the material and the reference one, respectively. f_{ref} is tortuosity factor which depends on the water content and porosity.

$$P_c = -\frac{\rho_w RT}{M_W} \ln \left(\alpha_c + \frac{1}{2} \sqrt{\beta_c + \gamma_c \frac{\varepsilon_0}{w}} \right) \tag{8.41}$$

where ρ_w and M_w are the density and molar mass of water, respectively. α_c, β_c, γ_c are the parameters which depend on the material.

Nguyen and Amiri (2014, 2016) developed a model to investigate the effect of EDL on unsaturated concrete subjected to chloride attack. In this model, the transports of ions and humidity were combined with the EDL effect at macroscopic scale by using Poisson–Boltzmann equation. It was found that EDL has significant effect on the chloride transport.

8.3.2 Probabilistic Model

Bastidas-Arteaga (2011) proposed a comprehensive probabilistic model for chloride transport in unsaturated concrete. The uncertainties related to the chloride penetration are also considered by using random variables to represent the model's parameters and the materials' properties, such as chloride diffusion coefficient, density of concrete, and specific heat capacity. Environmental actions were also viewed as stochastic processes, as shown in Figure 8.14. This probabilistic model accounts for: (1) chloride binding capacity; (2) time-variant nature of temperature, humidity, and surface chloride concentration; (3) concrete aging; and (4) chloride flow in unsaturated conditions.

In the model, the governing equations were established on three phenomena: (1) chloride transport; (2). moisture transport; and (3) heat transfer. Each phenomenon is represented by a partial differential equation (PDE), and their interaction is considered by solving simultaneously the system of PDEs.

(1) Chloride transport

$$\frac{\partial C_{tC}}{\partial t} = \underbrace{\mathrm{div}\left(D_c \omega_e \vec{\nabla}\left(C_{fc}\right)\right)}_{\text{diffusion}} + \underbrace{\mathrm{div}\left(D_h \omega_e C_{fc} \vec{\nabla}\left(h\right)\right)}_{\text{convection}} \tag{8.42}$$

$$C_{tc} = C_{bc} + \omega_e C_{fc} \tag{8.43}$$

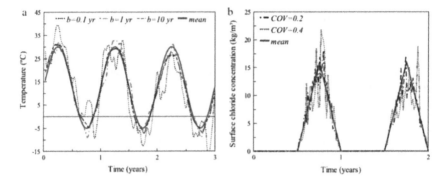

Figure 8.14 Realizations of (a) temperature and (b) surface chloride concentration. (From Bastidas-Arteaga et al. 2011.)

where C_{tc} is the total chloride concentration, t is the time, D_c is effective chloride diffusion coefficient, w_e is the evaporable water content, C_{fc} is the concentration of chlorides dissolved in the pore solution—i.e., free chlorides—D_h is the effective humidity diffusion coefficient, and h is the relative humidity.

(2) Moisture transfer

$$\frac{\partial \omega_e}{\partial t} = \frac{\omega_e}{\partial h}\frac{\partial h}{\partial t} = \text{div}\left(D_h \vec{\nabla}(h)\right) \tag{8.44}$$

D_h depends on the relative humidity in pore, temperature, and degree of hydration:

$$D_h = D_{h,ref} g_1(h) g_2(T) g_3(t_e) \tag{8.45}$$

The function $g_1(h)$ takes into consideration the dependence on pore relative humidity of concrete: The function $g_2(T)$ accounts for the influence of temperature on Dh $g_3(te)$, considers the dependence on the degree of hydration attained in concrete.

(3) Heat transfer

$$\rho_c c_q \frac{\partial T}{\partial t} = \text{div}\left(\lambda \vec{\nabla}(T)\right) \tag{8.46}$$

where ρ_c is the density of concrete, c_q is the concrete's specific heat capacity, λ is the thermal conductivity of concrete and T is the temperature inside the concrete matrix at time t.

Table 8.4 Probabilistic Models of the Random Variables

Variable	Mean	COV	Distribution
C_{th}	0.5wt%cement	0.20	Normal
C_t	40 mm	0.25	Normal(truncated at 10 mm)
$D_{b,ref}$	3×10^{-10} m/s^2	0.20	Log-normal
α_0	0.05	0.20	Beta over[0.025;0.1]
n	11	0.10	Beta over[6;16]
$D_{c,ref}$	3×10^{-11} m/s^2	0.20	Log-normal
m	0.15	0.30	Beta over[0;1]
U_c	41.8 kJ/mol	0.10	Beta over[32;44.6]
ρ_c	2400 kg/m^3	0.04	Normal
λ	2.5W/(m°C)	0.20	Beta over[1.4;3.6]
C_q	1000J/(kg°C)	0.10	Beta over[840;1170]

This model accounts for the most factors affecting chloride penetration into concrete in service, as shown in Table 8.4. There still are many areas in which further research is needed to improve the probabilistic model of chloride penetration, as pointed out in the following:

- Determination of model parameters for a wide range of concrete types
- Formulation and implementation of a model that considers the kinematics between concrete cracking and chloride penetration
- Study of the influence of hourly, daily, and weekly variations of temperature and humidity on chloride ingress
- Assessment and consideration of the correlation of material properties and climatic conditions from experimental data
- Consideration of the spatial variability of the phenomenon
- Characterization and modeling of error propagation in the whole deterioration process

Pradelle et al. (2016) gave a detailed comparison among some existing models for the prediction of chloride transport in concrete. They found that the selected models provide satisfactory predictions only in the case of concretes fully immersed in a saline solution. In case of materials in tidal zone, improvement of the models is needed to account for the effects of wetting–drying cycles on chloride ingress and/or dissolution–precipitation phenomena of hydration products, and carbonation.

8.4 CHLORIDE-RELATED DURABILITY CODES

It is well known that concrete in reality suffers from various durability issues, such as sulfate attack, freezing–thawing cycles, carbonation, alkali-aggregate reaction, and physical abrasion. In most cases, coupled deterioration modes act synergistically on concrete structures. This makes the prediction of concrete durability more complicated. Thus, holistic concept should be used to make durable concrete. The use of good materials, proper mixture proportioning, appropriate placement practices, and workmanship are essential to the production of durable concrete. However, it is not the intention to discuss the whole durability of concrete. This beyond the scope of this book. Only chloride-related codes of concrete structure are discussed in this book. The chloride related regulations from durability codes of the United States, China, Europe, and Japan are discussed briefly.

8.4.1 ACI Standards

There are several ACI standards/codes dealing with the durability of concrete. The major ones are listed as follows:

- ACI 201.2R-16 Guide to Durable Concrete
- ACI 222.3R-02 Guide to Design and Construction Practices to Mitigate Corrosion of Reinforcement in Concrete Structures
- ACI 318-11 Building Code Requirements for Structural Concrete

In ACI 201.2R-16 and ACI 222.3R-02, some general principles for corrosion-related properties of concreting materials are given. The pros and cons of using different materials were given, instead of specific requirements.

In ACI 318-11, specific requirements are given on the concreting materials and structural design. According to the detrimental sources, the environmental actions are classified into several categories, among which two categories are related with chloride attack. Exposure F refers to exterior concrete exposed to moisture and cycles of freezing and thawing, with or without de-icing chemicals. Exposure C necessitates additional protection against corrosion of steel of reinforced or prestressed concrete. Several classes based on the intensity of environmental actions are specified for each exposure. The chloride-related durability requirements are given in terms of the concrete strength class and the maximum water-to-cementitious materials ratio, as shown in Table 8.5. The additional requirements on the properties of concrete are specified for concrete subjected to freezing–thawing attacks.

8.4.2 Eurocode

Eurocodes provide a comprehensive framework for durability design for concrete structure design on the material level and structural level. This

Table 8.5 Environmental Classification and Durability Requirements for Structural Concretes by ACI-318 Code

Class	Conditions	Minimum Strength (psi)ᵃ	Maximum w/cm	Cementitious Binder	Maximum Cl⁻ Content (% cm)
C0	Dry concrete, or protected from moisture	2500	–	–	1.00 (0.06)
C1	Exposed to moisture, without external chlorides	2500	–	–	0.30 (0.06)
C2	Exposed to moisture and external chlorides	5000	0.40	–	0.15 (0.06)

framework considers the environmental actions and the design service life of structures.

- Material level: EN206–1, Concrete–Part1: Specification, Performance, Production and Conformity (CEN, 2000)
- Structure level: EN 1992–1–1, Design of Concrete Structures: General Rules and Rules for Buildings (CEN, 2004)

In EN206–1, the intensity of environment actions are categorized into different classes. The requirements for materials subjected to different exposure conditions are formulated by the prescriptive approach. The strength class is limited, the w/c ratio and cement content are also limited, as shown in Table 8.6. On structural level, the concrete cover depth and cracking width are also limited. Alternatively, Eurocodes also give performance-related approaches which consider the type and form of structure, local environmental conditions, level of execution, and the design service life.

Table 8.6 Chloride-Related Environmental Classification and Durability Requirements from EN206-1 and EN1992-1-1

Mechanism	Class	Exposure	Mini. Strength Class	Max. W/C	Mini. Cement Content (kg/m³)	Mini. Cover Thickness
Chloride/ deicing (XD)	XD1	Moderate RH	C37	0.55	300	30
	XD2	Wet	C37	0.55	300	35
	XD3	Dry-wet cycle	C45	0.50	320	40
Chloride/ Searwater (XS)	XS1	Airborne salt	C37	0.50	300	35
	XS2	Submerged	C45	0.45	320	40
	XS3	Tidal, splashing	C45	0.45	340	45

8.4.3 Chinese Code

Due to the massive infrastructure in China, different government departments propose different codes for durability design of concrete structures for various industry sectors, as follows:

- DLT 5241-2010 Technical Specifications for Durability of Hydraulic Structures
- DBJ 43/T-2014 Technical Specifications of Concrete Durability for Underground Construction
- TB1005-2010 Code for Durability Design on Concrete Structure for Railway Concrete Structures

The above-mentioned codes are based on a national code:

- GB/T 50476-2008, Code for Durability Design of Concrete Structures

In this chapter, GB/T 50476-2008 is mainly discussed. Generally, the code GB/T 50476 follows a prescriptive method. Like Eurocode, the exposure conditions are first classified into several classes: carbonation-induced reinforcement corrosion, chloride-induced reinforced corrosion in marine environment or de-icing salt environment, freeze–thaw cycle attack, chemical attack. But the detailed classifications of exposure conditions are different between China code and Eurocode. Chloride-related exposure conditions are as follows:

- Marine exposure (Class III) refers to the conditions for corrosion of reinforced steel in concrete induced by the ingress of chlorides in a marine environment.
- De-icing and other salts' exposure (Class IV) refers to the conditions for corrosion of reinforced steel in concrete induced by the ingress of chlorides in de-icing salt or other salts, instead of marine environment.

The code GB/T 50476 grades the intensity of the deterioration actions from A to F, with A standing for the slightest action intensity and F for the most severe. The deterioration intensity of a marine environment is graded into III—C, D, E, and F—and the deterioration intensity of de-icing and other salts into IV—C, D, and E—as shown in Table 8.7.

For different intended service lives, the concrete strength grade, the w/c ratio, and the concrete cover thickness are specified in the codes, as shown in Table 8.8. For important structures, the requirements for chloride resistance of concrete is also presented, as shown in Table 8.9. The chloride resistance is measured according to NT Build 492.

8.4.4 Japanese Code

Japanese codes provide the general requirements and standard methods for design of concrete structures.

- Standard specification for Concrete Structures – 2007 [Design]

Table 8.7 Chloride-Related Environmental Classification in GB/T50476-2008

Class	Environment	Intensity	Exposure Condition
III	Marine	C	Immersion in seawater
		D	Exposed to marine air and airborne salts
		E	Heavy salty air, tidal, and splash zones in mild climate
		F	Tidal and splashing zones in hot marine climate
IV	Deicing and other salts	C	Slight deicing salt fog; immersion in chloride water; contact with water with low chloride content and drying–wetting cycles
		D	Deicing salt spray; water with medium chloride content and drying–wetting cycles
		E	Direct contact with deicing salts; heavy spray of deicing salts; water with high chloride content and drying–wetting cycles

The Japan code adopts mainly a performance-based approach for durability design. Prescriptive codes are also provided for concrete structures subjected to carbonation.

Japan code classifies the exposure conditions into two main categories: reinforcement corrosion and concrete deterioration. For the reinforcement corrosion, three criteria are used to evaluate the durability limit:

- Crack width beyond the limit or not
- Chloride content at the surface of steel bar beyond the limit or not
- Carbonation at the steel bar beyond the limit or not

The requirements for concrete durability due to reinforcement corrosion are given in Table 8.10.

8.5 SUMMARY

Many efforts have been devoted to establishing models to predict the service life of concrete structures. It is a basis for a decision maker or investor to make a decision on investing a large amount of money on a new infrastructure and the maintenance strategy of existing infrastructure. Many models have been developed on the prediction of chloride transport in concrete, including the deterministic model and probabilistic model. The deterministic model includes the empirical model and physical model. The empirical models are often based on Fick's law; middle-term experimental data were implemented in the model, and some key parameters can be obtained. Thus, the long-term experimental data can be predicted from the empirical model. The accuracy of this kind of model depends on the key

Table 8.8 Requirements for Cover Thickness (mm), w/c and Strength Grade (MPa) of Concrete Subjected to Chloride Environment

	100 a			50 a			30 a		
	Strength Grade	Max. w/c Ratio	Cover Thickness	Strength Grade	Max. w/c Ratio	Cover Thickness	Strength Grade	Max. w/c Ratio	Cover Thickness
III-C,IV-C	C45	0.40	45	C40	0.42	40	C40	0.42	35
III-D,IV-D	C45	0.40	55	C40	0.42	50	C40	0.42	45
	≥C50	0.36	50	≥C45	0.40	45	≥C45	0.40	40
III-E,IV-E	C50	0.36	60	C45	0.40	55	C45	0.40	45
	≥C55	0.36	55	≥C50	0.36	50	≥C50	0.36	40
III-F	≥C55	0.36	65	C50	0.36	60	C50	0.36	55
				≥C55	0.36	55			
III-C,IV-C	C45	0.40	50	C40	0.42	45	C40	0.42	40
III-D,IV-D	C45	0.40	60	C40	0.42	55	C40	0.42	50
	≥C50	0.36	55	≥C45	0.40	50	≥C45	0.40	40
III-E,IV-E	C50	0.36	65	C45	0.40	60	C45	0.40	50
	≥C55	0.36	60	≥C50	0.36	55	≥C50	0.36	45
III-F	C55	0.36	70	C50	0.36	65	C50	0.36	55
				≥C55	0.36	60			

Table 8.9 Requirement for Chloride Resistance of Concrete

Design Service Life	100 a		50 a	
Action intensity	D	E	D	E
28d D_{RCM} (10^{-12} m²/s)	≤7	≤4	≤10	≤6

Table 8.10 Durability Requirements (Partial) for Reinforcement Corrosion and Concrete Deterioration from JSCE Guidelines

			Minimum	Performance Requirements	
Mechanism	Environment	Maximum w/cm	Concrete Cover (mm)	Crack Width[a]	Criterion
Reinforcement corrosion	Normal	0.50[b]/P[c]	40[b]/P	0.005c	Carbonation depth
	Corrosive	0.50[b]/P	40[b]/P	0.004c	Chloride content
	Severely corrosive	P	P	0.0035c	

[a] c is the thickness of concrete cover.
[b] Values for carbonation processes, beam elements and intended working life of 100 years.
[c] P designates the values determined by performance-based approach.
[d] Values not in parentheses are for general elements and values in parentheses are for thin elements.

parameters obtained from middle-term experimental data. And the application of this model may be limited to similar environment. The physical models take chemical–physical effects into consideration as scientifically correct as possible. Thus, it has wide application and solid scientific foundation. However, since the chloride transport process is too complicated, there is no generally accepted model which can precisely predict the transport of chloride in various situations.

Basically, the parameters for the chloride modeling are stochastic, and the probabilistic model might be a good option for predicting chloride transport. In a probabilistic model, the model parameters are treated as continuous stochastic variables, which are characterized by a mean value, a standard deviation, and a probabilistic density function. However, the developed probabilistic model is still far from practical.

It is worth mentioning that many of the models are developed on saturated concrete, since this phenomenon is the simplest situation. More efforts should be devoted to the unsaturated concrete.

The prediction of service life of concrete in chloride-bearing environments still needs extensive research, because the methods of calculation and the tests are far from satisfactory and have not been yet validated over the long term. All the results should be taken with precaution and used more as a rational and expert engineering assessment than a mathematical precise calculation. The durability codes standardised by different countries have been used in many key infrastructures as a durability design tool in a harsh service environment. Whether the durability of the infrastructures perform as designed is still a question.

REFERENCES

Ababneh A, Benboudjema F, Xi Y. Chloride penetration in nonsaturated concrete. *Journal of Materials in Civil Engineering.* 2003, 15:183–191.

Amey SL, Johnson DA, Miltenberger MA. Predicting the service life of concrete marine structures: An environmental methodology. *ACI Structural Journal.* 1998, 95(1):27–36.

Alexander M, Thomas M. Service life prediction and performance testing—Current developments and practical applications. *Cement and Concrete Research.* 2015, 78, Part A:155–164.

Altmann F, Mechtcherine V. Durability design strategies for new cementitious materials. *Cement and Concrete Research.* 2013, 54:114–125.

Angst UM. Predicting the time to corrosion initiation in reinforced concrete structures exposed to chlorides. *Cement and Concrete Research.* 2019, 115:559–567.

Arsenault J, Gigas JP, Ollivier JP. Determination of chloride diffusion coefficient using two different steady-state methods: Influence of concentration gradient. In: Nilsson LO and Ollivier JP (eds.). Proceedings of the International RILEM Workshop. St. Remy les Chevreuse, France, 1995, 150–160.

Bastidas-Arteag E, Chateauneuf A, Sánchez-Silva M, Bressolette Ph, Schoefs F. A comprehensive probabilistic model of chloride ingress in unsaturated concrete. *Engineering Structures.* 2011, 33:720–730.

Boddy A, Bentz E, Thomas MDA, Hooton RD. An overview and sensitivity study of a multimechanistic chloride transport model—Effect of fly ash and slag. *Cement and Concrete Research.* 1999, 29:827–837.

Cao Y, Gehlen C, Angst UM, Wang L, Yao Y. Critical chloride content in reinforced concrete—An updated review considering Chinese experience. *Cement and Concrete Research.* 2019, 117:58–68.

Crank J. *The Mathematics of Diffusion*, 2nd ed. Oxford University Press, London; 1975.

Dura. Crete, Modelling of Degradation, EU-Project (Brite EuRam III) No. BE95-1347, Probabilistic performance-based durability design of concrete structures, 1998. Report, vol. 4–5.

Fenaux M, Reyes E, Gálvez JC, Moragues A. Modelling the transport of chloride and other ions in cement-based materials. *Cement and Concrete Composites.* 2019, V97:33–42.

Goto S, Roy DM. Diffusion of ions through hardened cement pastes. *Cement and Concrete Research.* 1981, 11:751–757.

Gjørv OE. Durability design and quality assurance of major concrete infrastructure. *Advances in Concrete Construction.* 2013, 1:45–63.

Gulikers J. Probabilistic service life modeling of concrete structures: Improvement or unrealistic? Concrete under Severe Conditions: Environment and Loading, Tours, France, 2007, 891–902.

Hussainm SE, Rasheeduzzafar AM, Al-Gahtani AS. Factors affecting threshold chloride for reinforcement corrosion in concrete. *Cement and Concrete Research.* 1995, 25(7):1543–1555.

Johannesson B, Yamada K, Nilsson LO, Hosokawa Y. Multi-species ionic diffusion in concrete with account to interaction between ions in the pore solution and the cement hydrates. *Materials and Structures.* 2007, 40(7):651–665.

Justnes H., Kim MO, Ng S, Qian X. Methodology of calculating required chloride diffusion coefficient for intended service life as function of concrete cover in reinforced marine structures. *Cement and Concrete Composites.* 2016, V73:316–323.

Kassir MK, Ghosn M. Chloride-induced corrosion of reinforced concrete bridge decks. *Cement and Concrete Research.* 2002, 32(1):139–143.

Khitab A, Lorente S, Ollivier JP. Predictive model for chloride penetration through concrete. *Magazine of Concrete Research.* 2005, 57:511–520.

Kim J, McCarter WJ, Suryanto B, Nanukuttan S, Basheer PAM, Chrisp TM. Chloride ingress into marine exposed concrete: A comparison of empirical- and physically-based models. *Cement and Concrete Composites.* 2016, 72:133–145.

Kirkpatrick TJ, Weyers RE, Anderson-Cook CM, Sprinkel MM. Probabilistic model for the chloride-induced corrosion service life of bridge decks. *Cement and Concrete Research.* 2002, 32:1943–1960.

Li K. *Durability Design of Concrete Structures: Phenomena, Modeling, and Practice.* John Wiley & Sons; 2016.

Li KF, Zhang DD, Li QW, Fan ZD. Durability for concrete structures in marine environments of HZM project: Design, assessment and beyond. *Cement and Concrete Research.* 2019, V115:545–558.

Lindvall A. Probabilistic performance based life time design of concrete structures—Environmental actions and response. International Conference on Ion and Mass Transport in Cement-based Materials, Toronto, 1999.

Lindvall A. A probabilistic, performance based service life design of concrete structures-environmental actions and response. 3rd International PhD Symposium in Civil Engineering, Wien, 5–7 October 2002, 1–11.

Maage M, Helland S, Poulsen E, Vennesland Ø, Carlsen JE. Service life prediction of existing concrete structures exposed to marine environment. *ACI Materials Journal.* 1996:602–608.

Maaddawy El, Soudki TK. A model for prediction of time from corrosion initiation to corrosion cracking. *Cement and Concrete Composites* 2007, 29(3):168–175.

Mangat PS, Molloy BT. Predicting of long term chloride concentration in concrete. *Materials and Structure.* 1994, 27:338–346.

Marchand J. Modeling the behavior of unsaturated cement systems exposed to aggressive chemical environments. *Materials and Structures.* 2001, 34:195–200.

Martys NS. Diffusion in partially-saturated porous materials. *Materials and Structures.* 1999, 32:555–562.

Masi M, Colella D, Radaelli G, Bertolini L. Simulation of chloride penetration in cement-based materials. *Cement and Concrete Research.* 1997, 27:1591–1601.

Nguyen PT, Amiri O. Study of electrical double layer effect on chloride transport in unsaturated concrete. *Construction and Building Materials.* 2014, 50:492–498.

Nguyen PT, Amiri O. Study of the chloride transport in unsaturated concrete: Highlighting of electrical double layer, temperature and hysteresis effects. *Construction and Building Materials*. 2016, 122:284–293.

Nguyen TQ, Baroghel-Bouny V, Dangla P. Prediction of chloride ingress into saturated concrete on the basis of a multi-species model by numerical calculations. *Computers and Concrete*. 2006, 3:401–422.

Nokken MT, Boddy A, Hooton RD, Thomas MDA. Time dependent diffusion in concrete—Three laboratory studies. *Cement and Concrete Research*. 2006, 36:200–207.

Pang L, Li Q. Service life prediction of RC structures in marine environment using long term chloride ingress data: Comparison between exposure trials and real structure surveys, *Construction and Building Materials*. 2016, 113:979–987.

Petcherdchoo A. Closed-form solutions for modeling chloride transport in unsaturated concrete under wet-dry cycles of chloride attack. *Construction and Building Materials*. 2018, 176:638–651.

Reardon EJ. Problems and approaches to the prediction of the chemical composition in cement/water systems. *Waste Management*. 1992, 12:221–239.

Pradelle S, Thiéry M, Baroghel-Bouny V. Comparison of existing chloride ingress models within concretes exposed to seawater. *Material and Structure*. 2016, 49:4497–4516.

Saetta A, Scotta R, Vitaliani R. Analysis of chloride diffusion into partially saturated concrete. *ACI Materials Journal*. 1993, 5:441–451.

Samson E, Lemaire G, Marchand J, Beaudoin J. Modeling chemical activity effects in strong ionic solutions. *Computational Materials Science*. 1999, 15:285–294.

Samson E, Marchand J. Modeling the effect of temperature on ionic transport in cementitious materials. *Cement and Concrete Research*. 2007a, 37:455–468.

Samson E, Marchand J. Modeling the transport of ions in unsaturated cement-based materials. *Computers and Structures*. 2007b, 85:1740–1756.

Samson E, Marchand J, Beaudoin JJ. Modeling the influence of chemical reactions on the mechanisms of ionic transport in porous materials—An overview. *Cement and Concrete Research*. 2000, 30:1895–1902.

Samson E, Marchand J, Snyder KA. Calculation of ionic diffusion coefficients on the basis of migration test results. *Materials and Structure*. 2003, 36:156–165.

Samindi SM, Samarakoon MK, Sælensminde J. Condition assessment of reinforced concrete structures subject to chloride ingress: A case study of updating the model prediction considering inspection data. *Cement and Concrete Composites*. 2015, 60:92–98.

Schmidt F, Rostasy FS. A method of calculation of the chemical composition of the concrete pore solution. *Cement and Concrete Research*. 1993, 23:1159–1168.

Srubar WV. Stochastic service-life modeling of chloride-induced corrosion in recycled-aggregate concrete. *Cement and Concrete Composites*. 2015, V55:103–111.

Stanish K, Thomas M. The use of bulk diffusion tests to establish time dependent concrete chloride diffusion coefficients. *Cement and Concrete Research*. 2003, 33:55–62.

Shakouri M, Trejo D. A time-variant model of surface chloride build-up for improved service life predictions. *Cement and Concrete Composites*. 2017, V84:99–110.

Shakouri M, Trejo D. A study of the factors affecting the surface chloride maximum phenomenon in submerged concrete samples. *Cement and Concrete Composites*. 2018, V94:181–190.

Tang L. Chloride transport in concrete—Measurement and prediction. Ph.D. thesis, Department of Building Materials, Chalmers University of Technology, Goteborg, Sweden; 1996.

Tang L. Concentration dependence of diffusion and migration of chloride ions. Part 1. Theoretical considerations. *Cement and Concrete Research*. 1999, 29:1463–1468.

Tang L, Lars-Olof N. Modeling of chloride penetration into concrete—Tracing five years field exposure. *Concrete Science and Engineering*. 2000, 2:170–175.

Tang L. Engineering expression of the ClinConc model for prediction of free and total chloride ingress in submerged marine concrete. *Cement and Concrete Research*. 2008, 38:1092–1097.

Tang L, Nilsson LO, Basheer M. *Resistance of Concrete to Chloride Ingress: Testing and Modelling*. Taylor & Francis Group, LLC; 2012.

Takewaka K, Mastumoto S. Quality and cover thickness of concrete based on the estimation of chloride penetration in marine environments. In: Malhotra VM (ed.). Proceedings of 2nd International Conference Concrete in Marine Environment, ACI SP-109, 1988, 381–400.

Tuutti K. *Corrosion of Steel in Concrete*. Swedish Cement and Concrete Research Institute, Stochkolm; 1982, 486p.

Thomas M, Bamforth PB. Modelling chloride diffusion in concrete: Effect of fly ash and slag, *Cement and Concrete Research*. 1999, 29:487–495.

Tran V, Soive A, Baroghel-Bouny V. Modelisation of chloride reactive transport in concrete including thermodynamic equilibrium, kinetic control and surface complexation. *Cement and Concrete Research*. 2018, 110:70–85.

Truc O. Prediction of chloride penetration into saturated concrete—Multi-species approach. Ph.D. thesis, Deparment of Building Materials, Chalmers University of Technology, Goteborg, Sweden; 2000.

Weyers RE. Service life model for concrete structure in chloride laden environments. *ACI Materials Journal*. 1998:445–453.

Wiens U, Breit W, Schiessl P. Influence of high silica fume and high fly ash contents on alkalinity of pore solution and protection of steel against corrosion. Proceedings of the 5th International Conference on the Use of Fly Ash, Silica Fume, Slag and Natural Pozzolan in Concrete, SP-153, American Concrete Institute, Milwaukee, WI, 1995, Vol. 2, 741–762.

Williamson GS, Weyers RE, Brown MC, Sprinkel M. Validation of probability-based chloride-induced corrosion service-life model. *Aci Materials Journal*. 2008, 105(4):375–380.

Yu H. Study on high performance concrete in salt lake: Durability, mechanism and service life prediction. Ph.D. thesis. Department of Materials Science and Engineering, Southeast University, Nanjing, China; 2004.

Yuan Q. Fundamental studies on test methods for the transport of chloride ions in cementitious materials. Ph.D. thesis. Department of Structural Engineering, Ghent University, Ghent, Belgium; 2009.

Yuan Q, Deng D, Shi C, De Schutter G. Chloride binding isotherm from migration and diffusion tests. Journal of Wuhan University of Technology-Material Science Ed. 2013, 1–9.

Yuan Q, Shi C, Schutter G, Deng D, He F. Numerical model for chloride penetration into saturated concrete. *Journal of Materials in Civil Engineering.* 2011, 23:305–311.

Zhang J-Z, Buenfeld NR. Presence of possible implications of membrane potential in concrete exposed to chloride solution. *Concrete Science and Engineering.* 1997, 27:853–859.

Zhang M, Ye G, van Breugel K. Modelling of ionic diffusivity in non-saturated cement-based materials using lattice Boltzmann method. *Cemment and Concrete Research.* 2012, 42:1524–1533.

Zhang T. Chloride diffusivity in concrete and its measurement from steady-state migration testing. Ph.D. thesis. Department of Building Materials, Norwegian University of Science and Technology, Trondheim, Norway; 1997.

Zhang Y, Zhang M. Transport properties in unsaturated cement-based materials—A review. *Construction and Building Materials.* 2014, 72:367–379.

Index

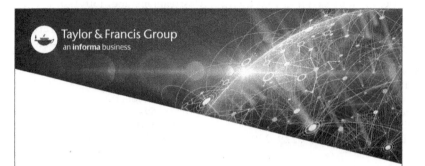